For some time now there has been a great deal of concerned reflection on how the tropical lowlands of the Americas have been perceived and exploited. This book addresses that concern. It is also something of an appreciation of tropical lowlands, as they emerge along one particular road, gathered from the accounts of early nineteenth-century observers.

Aerial reconnaissance has shown that many wetlands in the lowlands which these travellers crossed are patterned with the remains of Prehispanic platforms and canals, an old and effective system for the cultivation of wetlands. These show particularly clearly in the pastures of modern ranches—a very different land use, and yet perhaps governed by similar constraints. The pastures are dotted with palms which eloquently indicate repeated burning and long use and are scored by drainage ditches cut according to contemporary practice, thus giving evidence of both ancient and modern use. The travellers' accounts throw light on this juxtaposition.

Early nineteenth-century visitors to Mexico usually entered the country at Veracruz and proceeded inland along the Jalapa road. They produced a rich literature that reveals much about the region as well as what the European and North American travellers thought about the tropics.

The reader is taken along the Veracruz-Jalapa road up to the summit of the pass and on to the central tableland and allowed to see the coastal landscape take shape from the commentary, step by step—detailed and coloured by predisposition, the "objective" landscape often aggrandized and misperceived. The accounts are not benign; they are tinged with an evaluation of tropical lowlands that unfortunately persisted and proved prejudicial to actual development here and elsewhere.

In this book, Alfred H. Siemens allows a wide array of commentary to coalesce as though it were a piece of landscape theatre—fascinating and at times entertaining, but also carrying venom.

ALFRED H. SIEMENS is a professor in the Department of Geography at the University of British Columbia.

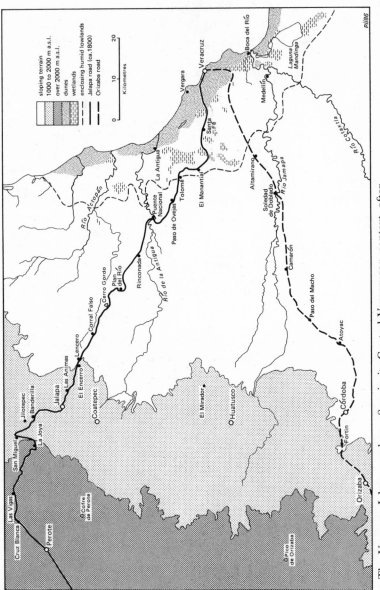

1 The Veracruz-Jalapa road ca. 1800, in its Central Veracruzan context, after Humboldt (1979/1811:9), field checks, and various additional sources

Map legend:

sloping terrain
1000 to 2000 m a.s.l.
over 2000 m a.s.l.
dunes
wetlands
enclosing humid lowlands
Jalapa road (ca.1800)
Orizaba road

Kilometres
0   10   20

Place names on map:

Boca del Río
Laguna Mandinga
Río Cotaxtla
Veracruz
Medellín
Vergara
Santa Fé
Río Actopán
La Antigua
Altamirano
Río Jamapa
Puente Nacional
Soledad de Doblado
Paso de Ovejas
Tolomé
El Menantial
Río de la Antigua
Rinconada
Plan del Río
Cerro Gordo
Corral Falso
Camarón
Lencero
Paso del Macho
El Encerro
Las Ánimas
Coatepec
El Mirador
Huatusco
Atoyac
Córdoba
Banderilla
Jalapa
Jilotepec
Fortín
La Joya
San Miguel
Orizaba
Las Vigas
Cofre de Perote
Cruz Blanca
Perote
Pico de Orizaba

pli86

Between the Summit and the Sea

Central Veracruz in the Nineteenth Century

Alfred H. Siemens

University of British Columbia Press
Vancouver 1990

ISBN 0-7748-0354-1

Canadian Cataloguing in Publication Data
Siemens, Alfred H., 1932–
Between the summit and the sea

Includes bibliographical references
ISBN 0-7748-0354-1

1. Veracruz (Mexico: State) – Description
and travel. 2. Mexico – Foreign public
opinion – History. I. Title.

F1371.S53 1990   972'.62'04   C90-091165-4

This book has been published with the help of a grant from the Social Science
Federation of Canada, using funds provided by the Humanities and Social
Sciences Research Council of Canada.

UBC Press
6344 Memorial Rd
Vancouver, BC V6T 1W5

For Alice

who accompanied me on this road and many others

Contents

Diplomats, representatives of investment and trading companies, merchants, naturalists, professional
writers, American soldiers, too, for a short time, and
adventurous travellers of other sorts, responded
avidly to a country made accessible by independence. They found the time an *época negra* and the
lowlands a backwater, but their portrait of them is a
fanciful tapestry, even though many authors maintained they were only presenting straightforward
facts.

The sighting of Mt. Orizaba animated many a ship's
passengers; before disembarkation they were well
into a characterization of the lowlands and their
inhabitants. They had also begun their inner struggle with the hazards of this entry into the tropics.

Veracruz was found to be clean and neat, but it also fairly glowed with the danger of death. During uneasy stays, often perceived as detention, the visitors described an alien urban structure, strange goings-on in the streets, and a population that in the main was not very impressive when compared with the town's foreign minority.

While in Veracruz, travellers tried to come to terms with the hazard that had overshadowed, and guarded, this coast from early colonial times, that affected the perceptions of all comers, and stymied the reflections of even the most thoughtful: the danger of death by yellow fever.

A rigorous journey lay ahead; highway robbery was common, at least much discussed. One had to arm oneself to travel, which reflected an attitude, not only in the face of a particular hazard but toward the country as a whole.

In the humid lowlands, just inland from the dunes, the travellers finally found tropical luxuriance. A dyadic evaluation was taking shape: one view of tropical people, another of tropical nature. The first was negative, the second positive, but qualified.

Civilian authors, with one notable exception, disregarded much of the hill land between the humid lowlands and Jalapa. American soldiers moving inland in 1847 neglected most everything else, but were eloquent in their descriptions of this tract, especially its vegetation. The complementary perceptions invoke the ecological complementarity of the regions themselves.

The visitors countered what they saw of tropical nature with scattered observations on tropical people.
Their roadside ethnography makes some distinctions, outlines a dominant subculture, and elabo-
rates on a basic indictment.

At about 1,000 meters above sea level the danger of
yellow fever abated and the air freshened. Everyone
rhapsodized with greater or lesser facility about paradise. In the process the two fundamental propositions regarding the tropical lowlands were modified,
but left essentially intact.

Above Jalapa the air cooled and the vegetation became coniferous. The road offered a last view of the
entire coastal landscape, which led many to summarize their journey thus far in terms of an altitudinal
construct. Then it was over the pass and on to Mexico City.

The visitors maintained that they yearned for facts,
and do provide much useful information, but they
also shaded what they saw in many rich, imaginative
tones. Unfortunately the portrait conveys an indictment; this can still be detected in the literature and
rhetoric on tropical lowlands in the second half of
the twentieth century and seems to have been prejudicial to the understanding and actual use of their
resources.

Tables and Figures

Preface

Most European and North American visitors to Mexico in the first half of the nineteenth century entered the newly independent country through the port of Veracruz and travelled along the Veracruz-Jalapa road—the main route into the interior during that time. The main objective of this book will be to allow the lowland landscape between the coast and the pass over the Sierra Madre Oriental to emerge from their accounts. They reconstruct an intriguing place at a critical time, but they also magnify what is, at least at first sight, rather unprepossessing terrain; they embellish and exaggerate it into a theatrical setting for the drama of their own journeys. This is all quite absorbing, even instructive, but for anyone who has spent time in the tropics this literature is also eventually unsettling. It is difficult not to notice the progressive elaboration of a set of preconceived ideas as the travellers proceed into the interior. One feels compelled to place their accounts into the light of recent thought on the nature of the tropics and on what should be done with them.

The book germinated unnoticed in dissertation research on mid-twentieth-century change in the settlement and agriculture of the lowlands of Southern Veracruz. Settlers were moving down from densely settled uplands, clearing the forests, and laying out new communities. Roads were being cut; ranching and commercial cropping were expanding. The rhetoric of the frontier could be heard in the new clearings and in the offices of the responsible governmental officials. It could be read in academic publications on the subject as well. Here in lowland Veracruz was movement toward the solution of at least some of Mexico's agrarian production problems; agrarian reform in the country at large had long since ground to a halt. This was a growth region; tropical lowlands were being made tractable, productive, and a source of exportable surpluses. The concern was how best to facilitate the process, how to minimize waste and disorder, especially how to avoid the costly blunders that plagued governmental projects. There was little doubt about the high potential of tropical lowland resources and the appropriateness of extra-tropical paradigms.

Many skeletons of the forest still stood in the fields and pastures. On maps and in the landscape itself it was clear that the new settlement was reaching out from older nodes. Questions of antecedents, of beginnings and directions, suggested themselves. It was possible to follow them up at two of the prime European collections of historical materials on Mexico: the Lingabibliothek at the University of Hamburg and the library of the Iberoamerikanisches Institut in West Berlin. Their ample holdings of accounts by nineteenth-century observers became a particular fascination. Supplemental materials were later obtained from the Library of Congress in Washington and other repositories. Out of it all emerged a tropical lowland backwater, its "development" still far in the future—a region that the observers had richly detailed but had also stigmatized.

As it happened, this inquiry was overshadowed in 1968 by another, which extended the reflection on human response to the tropical lowlands into a more distant past. During air reconnaissance over some newly established colonies in southwestern Campeche, several hundred kilometers to the east of the extremity of Veracruz, wetlands in the floodplain of the Candelaria River were found to be patterned with the rectilinear remains of prehistoric canals and planting platforms. Here finally was the evidence for the intensive Prehispanic Mayan agriculture that had been predicted; it would be easier now to explain the sustenance of the population concentrations implied by new archaeological data. Something similar eventually would become apparent in the wetlands of the lowlands of Central Veracruz. A number of extensive complexes of remains were recognized immediately adjacent to the road that most of the nineteenth-century observers took from the coast into the interior.

Several countervailing evaluations were now apparent. The nineteenth-century observers, beginning with Alexander von Humboldt, prescribed drainage for these wetlands. Development planners in the second half of the twentieth century, and landholders themselves, were thinking along the same lines. The only good wetland was a dry one. This was clearly very different from attitudes suggested by the prehistoric remains. What was interstitial in historic times had once been central. The wetlands had been favoured terrain, attracting settlement around their peripheries. An ancient indigeneous expedient obviously had been immensely successful.

The discussion of the new evidence repeatedly circled around to a logical but seemingly quite impractical consideration: would it not be possible to revive these practices in aid of improved food production in the lowlands? This paralleled the growing interest, particularly strong among Mexican students of the rural scene, in traditional agricultural systems as a means toward the solution of some persistent problems of marginalized peasants. Here were some possible alternatives to the problematical Green Revolution.

Various approaches to the resources of the tropical lowlands were thus juxtaposed; comparisons suggested themselves. The Prehispanic approach that could be inferred from the remains, and the ways of the traditional agriculturalist, still observable in various places within the country, contrasted with the terms of reference of nineteenth-century foreign observers. These, in turn, seemed familiar after my time spent in the field in Veracruz; they had certain affinities to the approach of recent in-migrants and the planners of modern development. The examination of the nineteenth-century accounts thus became a nexus for reflections on what had been and might yet be done with the lowlands. The concepts of "growth region" and "backwater," on which the reflections were originally based, had to be qualified. The sparse agriculture criticized by nineteenth-century observers became more understandable, extraneous paradigms seemed less and less appropriate, the "growth region" concept less appropriate.

The relationship between nineteenth-century evaluations of and twentieth-century approaches to tropical lowland development has been of particular interest in this context. To what extent were nineteenth-century evaluations prejudicial to twentieth-century tropical lowland development? The lineage of particular formulations, their diffusion, their possible causal connections—all these are intriguing issues, but difficult to trace and beyond the scope of this analysis. However, it is possible to test for the persistence of nineteenth-century evaluation into our times and to hypothesize an effect. It is also possible to see the gradual erosion of these inherited views.

These considerations provide one of the endpieces for the core of the analysis: the accounts themselves and the landscape they create. The other is the set of introductory considerations presented in Chapter 1: the foreign interests represented by the observers, who they were and what their observations were like as literature, the nature of the terrain they travelled over and something of the depressed economy and chaotic politics of the time. Some readers may well want to dispense with both the endpieces, at least initially, and join the travellers straightaway.

The volumes of the observers' accounts have been much anthologized but little analysed, certainly not with the focus on the tropical lowlands that is proposed here. Various broad, systematic studies have taken them into consideration, as will be outlined in Chapter 1, but they have not been read in quite the same way that we will read them and not allowed to coalesce to the same effect. The overall animadversion of foreign observers toward things Mexican has often been noted, of course, but here it shows what may be unexpected facets.

A common itinerary in the accounts provides the organization of the core of the book, and facets of a shared predisposition provide the themes for its individual chapters. There were common shifts in attention along the way—breaks in the perceptual slope, so to speak. These coincided to a considerable extent with the altitudinal zonation of physical environmental features. During the course of the common journey both the landscape and the predisposition progressively take shape for the reader. When those of the observers who travelled inland arrived at a particular lookout above Jalapa, they could, on a good day, look across the entire lowland and catch a glint of the sea. Here they might well become effusive and reflective; they might even attempt a synthesis. For many this was the first of a series of culminations in their journey into the New World.

There were artists among the visitors to Mexico and their images add to our enjoyment and understanding. They have been treated even more cavalierly than the anthologized accounts in the literature on the nineteenth century. The pictures are often poorly reproduced and inadequately cited. Those that treat lowland subjects have been sought out and analysed for their relationships to the literary imagery. A selection has been photographed and printed in high contrast, and in part redrawn, in order to freshen them. There are not many in all; the artists were busier elsewhere, particularly after they had arrived in Mexico City.

Various other kinds of illustrational material have been drawn in. A certain amount of cartography is indicated by a subject of this kind. Photographs of the contemporary cultural landscape have been avoided, in the main, in order not to set up superficial and digressive comparisons between then and now. But some evaluations through the viewfinder have been admitted, particularly where natural environmental features could be isolated from modern paraphernalia. Several air obliques are included too. They have often been resorted to throughout our investigations in the lowlands in order to achieve some perspective over flat landscapes. They surmount obstacles admirably, but are not yet views straight down, in which the features become symbols which one must learn to interpret.

The book remains primarily a literary analysis and the key to it is the reading across the grain, the bringing together, in turn, of what is said in all the accounts regarding each segment of the journey. This allows the observers to complement each other on a kind of matrix: the adventurer shading in what a diplomat or even a naturalist had missed at a particular point, all of them together elaborating a predisposition in response to the sequence of stimuli along the road.

Such a reading makes of the authors a company. Some of them were in fact personally acquainted, many more had heard of each other. They traded warnings and bits of advice; a few had given each other letters of recommendation. Almost all of them will have carried Humboldt in their baggage, literally or figuratively. His *Political Essay on the Kingdom of New Spain*, in its various editions, was widely known and influential throughout the time in which they made their observations. They professed related aims and could understand each other, given a minimum of translation. In one's mind they become a travelling seminar, noisy with affirmations and denunciations.

The eminent Mexican historian, Juan Antonio Ortega y Medina, went somewhat further. He took the travellers' observations quite seriously, all in all, but then at intervals he seems to have shaken his head over them. They were a motley, garrulous caravan of adventurers, or better still, a travelling theatre troupe (1955:2,19). Their accounts, indeed, are conjury—limited as sources of factual information, but most promising for an analysis of bias.

Acknowledgments

The bibliographical search that brought this company of travellers together began in the Lingabibliothek at the University of Hamburg and, on the other end of furtive dashes through an alien countryside, in the library of the Iberoamerikanisches Institut in West Berlin. Copies of a good number of the account were put in jeopardy one night by a border guard of the German Democratic Republic who insisted on unravelling the microfilm and examining it against a streetlight to see if it contained pornography. My thanks to the librarians at those two magnificent institutions, who helped me assemble the core of the material and copy it, as well as others at the Library of Congress, the library of Tulane University, and the library of my own University of British Columbia, who helped to enlarge the company of writers that was to be followed along the Veracruz-Jalapa road, and to obtain background material on them.

The analysis was made possible by several grants from the Social Sciences and Humanities Research Council of Canada, as well as smaller grants and several periods of leave from the University of British Columbia. The publication of this book has been facilitated by a subvention from the Social Science Federation of Canada.

Elizabeth Krupp helped in the early processing of the manuscripts. Various drafts were typed by Elizabeth MacInnes, Diana Phillips, and Irene Hull, who also assisted in early editing. Later versions were stored and revised at Jeeva's Word Processing. Paul Jantz, Angus Weller, and his associates, Grettchen Zdorkowsky and Robert Laird worked on the preparation of maps, photographs, and drawings. Patricia Luniw revised the references and proofread several versions of the text as well. I am very grateful to my colleague Richard Copley for his careful review of the chapter on yellow fever and to various anonymous pre-publication reviewers for their many suggestions. Susan Clark improved the final manuscript immensely with her thorough editing.

It was important from the beginning that this analysis should make sense to my Mexican friends. Several of them read the manuscript and put my mind at ease: I am indebted to Alba González Jácome, of the Universidad Iberoamericana, and Arturo Gómez-Pompa, director of the Instituto Nacional de Investigaciones sobre Recursos Bióticos. The book is a kind of thank you to Gastón Alvarez Fernández of Jalapa for his generosity and a most agreeable association of many years. It is also interest on an investment made years ago by an idealistic Mexican agronomist and patriot in the education of a Canadian graduate student. Leandro Molinar Meráz introduced me to Veracruz and merits my thanks, wherever he may be now.

Between the Summit and the Sea

CHAPTER ONE

The Travellers, the Lowlands, and the Times

One of the more assiduous visitors to newly independent Mexico some-how obtained an official count of foreigners entering the country as immigrants or travellers through its principal port of Veracruz in 1830: 99 French, 75 English, 53 Americans, 51 Germans, 13 Italians, 9 West Indians, 7 Swiss, 5 Central Americans, 2 Colombians, 2 Portuguese, 1 Hungarian, 1 Savoyard, and 59 of unknown origin, for a total of 377 (Koppe 1837:I, 154). Information like this on the chaotic early decades in the history of the new country is rare and must be cherished. It allows us to begin sketching in the background of the accounts that will be the basis of this analysis.

The numbers and nationalities will have varied from year to year, but the four nationalities most strongly represented in 1830 remained so throughout the first half of the century. Their languages dominate the foreign commentary available today on the Mexico of the time. The entire list indicates something of the range of foreign interest in the new republic and the many new possibilities it offered for investment, adventure and scientific inquiry.

We have access to the observations of only a small proportion of the foreigners who arrived. Of the 377 that came in 1830, for example, only about 1 per cent left published accounts. In some years it was more, but usually it was less. Fortunately, to paraphrase the introduction to one anthology of travel literature, a day in one journey is worth years in others (Flores Salinas 1964:10). The small sample is in fact a rich resource.

Who were these foreigners and what influenced their perspectives? What do their observations have in common and how do they comple-ment each other? These and related questions must be answered before the accounts themselves can be considered. One route into the interior organizes the accounts; the various elements of the lowland landscape cue a more or less coherent and cumulative sequence of thoughts about the tropics. Something systematic must therefore be said as well about that landscape, and about the time in which it was traversed. However, as has been noted, this is not so much a reconstruction of Central Veracruz during an interesting period in its history; for that a wider literature would be necessary. The imagery spun over this terrain is the main concern.

FOREIGN INTERESTS IN MEXICO AFTER INDEPENDENCE

The gist of the new country's external relations affected the composition of each year's "theatre-troupe" and, to some extent, what was seen along the way in the lowlands. One notices immediately that there are no Spaniards on the 1830 list of Carl Koppe, and one can understand why there was little enthusiasm on the peninsula for travel or new investment in the disaffected former colony. In 1821, after the attainment of Mexico's independence, all Spaniards in Mexico were declared citizens. However, in the following years, the official attitude toward them soured, largely because it was suspected that some were conspiring to facilitate Spain's reconquest of Mexico. In 1827 this ill feeling led to an expulsion of almost all Spaniards, a move that was unjust to the majority who had not conspired against the Mexican government, erosive of many family fortunes and impoverishing for Mexico itself since a great deal of capital was lost or taken out of the country. An ill-equipped Spanish expedition did land in Tampico in 1829, but it was repulsed.

Slowly a pro-Spanish reaction set in. By 1836 a treaty of peace and friendship had been concluded with the former mother country. The Spaniards who had been expelled were allowed to come back, and many did. Relations ran fairly smoothly for a time, but then Spain, together with Great Britain and France, pressed claims for compensation on grievances from before and after the winning of Independence. The joint expedition of 1862 and a great deal of bitterness were the result.

Spain's first envoy to the young republic was sent out in 1839. He was Don Ángel Calderón de la Barca, whose wife Frances eclipsed him for posterity with her account of life and travels in Mexico, a classic sometimes named in the same breath with Prescott's *History of the Conquest of Mexico* (1843). Madame Calderón de la Barca, born a Scot, and educated in Boston, occasionally assumes a Spanish perspective in that she describes her husband's mission in sympathetic terms, expresses her appreciation of the Spanish formalities still encountered in the former colony, and of the great works of the Conquistadores. However, she was also alienated by some aspects of the Mexican Spanish heritage. Mainly, this ascerbic lady represented her own sensibilities; the Spanish perspective on the new nation and its tropical uplands is hardly represented in the literature basic to this study.

Visitors from the emergent republics to the south and indeed close relationships between their countries and Mexico might have been expected given the commonality of the hemisphere's struggle for independence. The Federation of Gran Colombia sent an envoy to independent Mexico in 1822, but this representation was not sustained. By mid-century, there were consular relations with a number of other republics to the south but there was little trade for anyone to follow with a flag and, in any case, most governments were preoccupied with widespread internal unrest.

The British were the first Europeans to send diplomats; they were intensely interested in the economic prospects. They had deduced from Alexander von Humboldt's observations that Mexico was rich in resources and backward in technology—a perfect investment opportunity. The trade agreement between the two countries, taken through its various stages from 1825 to 1827, became a model for the agreements Mexico was to conclude with other European countries. Relations with Britain were generally peaceful and friendly throughout the first half of the nineteenth century, even though the initial ardour to invest cooled significantly and strong claims had to be made regarding unpaid debts. Britain was a participant in the joint European expedition of 1862 to force repayment.

German interest in Mexico was wide-ranging. It was well surveyed, together with the whole range of Mexico's foreign relations, in a thoughtful and scholarly, if somewhat bureaucratic manner by Baron Emil Karl Heinrich von Richthofen (1810–95), "minister-resident" of Prussia in Mexico at mid-century (1854:36–89). It seems a much neglected source in research on nineteenth-century Mexico. A Prussian and a diplomat himself, von Richthofen, as might have been expected, described diplomatic ties between the various German Länder and Mexico as in good order. Nor is this contradicted anywhere in our literature. There had been some claims by wronged Germans against Mexicans, but he considered the settlements reached to be honourable on both sides. The good relations initiated by Humboldt evidently were not dissipated for many decades. A substantial new collection of essays analyses the role of the Germans in post-colonial Mexico, and casts a rather less favourable light on the subject. Evidently this should be considered as imperialism too (von Mentz de Boege et al. 1982).

German entrepreneurs established substantial trading houses and industries. German capital and skills were engaged in the Mexican mines. Officials at home or already in Mexico debated whether or not

their countrymen should be encouraged to undertake agricultural settlement in Mexico. A predominantly negative opinion in this regard
took shape soon after Independence. Political conditions were too
unsettled and religious intolerance too great. Various Germans with a
scientific bent found Mexico interesting and came to continue Humboldt's inquiries into its geology, its plants, and other aspects of its
natural environment. German travel accounts are richer in these
respects than others. The German travellers observed economic life and
customs with a similar attention to detail. However, they were also convinced, one and all, that they had much to teach; denigration of the
Mexican was universal and varied only in degree.

France, long the source of ideas and manners for knowledgable Mexicans, entered into direct relations with Mexico relatively late, that is, in
the mid-1830s. These relations do not seem to have been as well
informed or thought out as they might have been. A recent study of the
French experience in Mexico corroborates the rather negative observations of Richthofen and points to a general nonchalance in the handling of Mexican affairs. There was no French Poinsett or Ward or
Richthofen. Official France was contemptuous of the Mexicans. They
were barbarous (Barker 1979:22–3,63). The picture that consular documents painted of the conditions in the new republic was certainly not
very prepossessing (Diaz 1974). Fortunately, most of the French authors
to be considered here were not too noticeably tinged by this contempt;
in fact, some of the most sensitive commentary on the Mexican tropical
landscape may be attributed to them, particularly Mathieu de Fossey,
Louis de Bellemare (Gabriel Ferry), Pharamond Blanchard, and Lucien
Biart.

Colonization played a role in the development of acrimonious
French-Mexican relations too. A French company concluded a contract
with the state of Veracruz to settle people of various nationalities on a
tract of land south of Coatzacoalcos. Several shiploads of prospective
settlers, mostly French and mostly very unprepared, were not given the
support promised by Mexican authorities and the French entrepreneurs. The whole venture failed miserably. When a German nobleman,
the Baron von Müller, crossed the Isthmus in 1857, only a few families
remained, all with stories of great hardship (1864:II,426–8).

Mexico reacted to French claims regarding unpaid debts and
wronged French citizens with an expulsion order in 1838 which
harmed many innocent people, as had the expulsion of the Spaniards
in 1827. In the same year a punitive French expedition under Admiral
Baudin appeared off the coast of Veracruz, blockaded the town, and for
a brief time held the fort of San Juan de Ulúa. However, by 1839 a peace
treaty had been signed and for a time relations improved, until old

claims were advanced again. In 1864 France took the lead in imposing the emperor Maximilian.

Mexico's volatile relations with the United States are a vast and much discussed subject. Throughout the accounts by American civil and military authors considered in this study runs the conviction of the inferiority of all things Mexican. The accounts of wronged American traders venturing south soon after Mexican Independence had had a wide circulation in the United States. American confidence and know-how, on the other hand, were daily evident along the Jalapa road in the form of American-built coaches driven at breakneck speeds by American drivers. Waddy Thompson, Esquire, "Late Envoy Extraordinary and Minister Plenipotentiary of the United States at Mexico," expressed his nation's expansiveness. He had arrived in Jalapa and was convinced that "No spot of earth will be more desirable than this for a residence whenever it is in the possession of our race, with the government and laws, which they carry with them wherever they go. The march of time is not more certain than that this will be, and probably at no distant day"(1846:13).

In 1847 the Americans defeated Santa Anna's forces at Cerro Gordo. Through it all the Americans drew both the Mexicans' admiration and resentment. The residue of the denigration on the one hand and the profound ambivalence on the other is still clearly evident in the news coverage of any modern encounter between the presidents of the two nations.

Much of the agitation one senses in the travellers' accounts and the logistics of many an individual journey stemmed from disturbances caused by the sequence of foreign military incursions that are not always easy to keep in order as one reads the individual accounts: the French demand for the satisfaction of various claims in 1837 followed by a blockade of Veracruz and the bombardment and capture of the fortress of San Juan de Ulúa; the American invasion at Anton Lizardo in 1847 and the subsequent battle of Cerro Gordo; the joint French, British, and Spanish landing in 1862, again to support claims for the payment of debts. The British and Spanish eventually withdrew, leaving the French to press their claims and to support the imposition of a European emperor, Maximilian. In addition, there was the troublesome situation in the port that delayed Independence, to which we will refer again, as well as skirmishes among rival Mexican forces that often affected traffic along the Jalapa road. One could be delayed by these events, or even blunder into shellfire. Some travellers went around the disturbances; entering and leaving through Alvarado to the south, for example, and thus inadvertently enriching their journals and our understanding of the country south of Veracruz.

Whatever the interests of the foreign visitor, they were usually focused on points in the interior. The mines, the governmental apparatus of the capital, and the most populous parts of the country were all up in the highlands. Most travellers hurried through the hot and humid lowlands, yet their first views of Mexico, and even more their first views of Mexicans, were freighted with not only their own but their country's predispositions. The indictments often emerge fully formed at the first encounter. However, one has the impression repeatedly that what was seen in the lowlands was seen by the way and described in an unguarded fashion. A diplomat sent to judge the potential for investment and trade, for example, would write with considerably more circumspection of the mining areas than of what was seen in the first days of his journey, along the way into the interior. Such unguarded commentary would seem particularly useful for an analysis of predisposition.

The visitor's actual confrontation with Mexico was likely to be dramatic. One thinks of the North American response to a more accessible China in the 1960s and 1970s. Indeed, Ortega y Medina, the distinguished Mexican historian, has developed the point that Mexico, with its isolation finally breached, appeared to many as a China of the New World (1955:37-42). It was little known, even with Humboldt's *Essay* in many hands. Its people had strange manners and did unheard-of things. Its world of plants and animals offered endless curiosities; its resources seemed fabulous. "In New Spain there is nothing either impossible or improbable" (Ward 1828:II, 23).

Actually entering Mexico and attempting to gain a foothold there was risky in the extreme. The wrecks of many ships garnished the beach to the north and south of the Veracruz wharf. One could easily die of yellow fever in the port itself. Duties were high, robbery was frequent, and justice throughout the country was arbitrary. All this left some sharp impressions.

To the Mexicans, of course, the foreigners were a spectacle. They were by turns welcomed, barely tolerated—particularly when it came to matters of religion—admired for the cut of their clothes, respected for their knowledge and skills, imitated, and always stared at. Crowds often gathered at the gates of the coach houses, even in the dead of night, to watch foreigners come and go. Undoubtedly there was a great deal of puzzled banter and a healthy quantum of derision. Various travellers noticed something more serious: an anti-foreign sentiment was growing, especially with each armed intervention. Literate Mexicans, of course, were sensitive to the way they were represented by foreigners. They were pained by much of what had been written about them and yet were convinced by much of it. Even Ortega y Medina, often ironic or sarcastic in his reference to the foreign visitors of the nineteenth century, shows an underlying respect for what they said (1955).

THE AUTHORS AND THEIR ACCOUNTS

The commentary of the "garrulous caravan of adventurers" may be divided into three broad categories: travel literature, military memoirs, which may be taken as a particular form of travel literature, and systematic sources which are in large part syntheses of repeated journeys that provide a factual matrix for the whole. The main concern in the choice of authors was the relevance of their material to the environs of the Veracruz-Jalapa road. However, since most foreign visitors to Mexico came or went via Veracruz, our sources include the majority of the names that would need to be cited in any other analysis of the early nineteenth century. The names and nationalities of the authors here included, as well as the dates and directions of their travel and the dates and titles of their publications, are presented in Table 1.

The authors are arranged according to the chronology of their observations, within the categories described. Read in this order, the accounts could be related easily to each other and to the major events of the period. The dates of first publication, frequently not parallel to those of observation, often clarified what influence one author had on another, or who might have been the source for an already familiar but unacknowledged idea. Each author had to be followed through at one sitting, from the Gulf to the mountain pass, or vice versa. In addition, they all had to be searched repeatedly for comparable commentary on any given feature—read both horizontally and vertically, as it were. Therefore the directions of the journeys had to be kept in mind. And sometimes the rendition of a feature varied with the direction from which it was seen.

This sort of literature does not easily cohere. As already noted, it has been more anthologized than analysed. It includes information on a great many different subjects and the ordering of it can easily become an exercise in categorization, as for example in the footnotes of Gardiner's book on the journal and correspondence of the American traveller Edward Thornton Tayloe (Gardiner 1959). Irving Leonard's very readable anthology of *Colonial Travellers in Latin America* (1972) exemplifies a related tendency. The pieces are extraordinary, full of curious information presented in a variety of styles, but the introduction is mainly background and the approach antiquarian. William Mayer presented *Early Travellers in Mexico* in 1961. Berta Flores Salinas went through essentially the same list in 1964, without reference to Mayer. Both affirm some basic historiographic imperatives: the need to understand the context of the journeys, to evaluate the accounts themselves for their

TABLE 1: Prime Sources

Travel literature						
Date of observation	Direction of travel		Name	Title	First publication	Nationality of author
	W	E				
1822	↓		Poinsett, Joel Roberts	Notes on Mexico	1824	American
1823	↓		Bullock, William	Six Months' Residence and Travel in Mexico	1825	English
1823		↑	Ward, H.G	Mexico in 1827	1828	English
1825	↓	↑	Tayloe, Edward	Mexico: 1825–1828 (Gardiner, ed.)	1959	American
1826 (?)	↓		Schiede, C.J.W.	"Berichte" in Linnaea	1829–30	German
1828	↓		Sealsfield, Charles	Der Virrey . . .	1834	German
1829 (?)		↑	Beltrami, J.C.	Le Mexique	1830	Italian
1830		↑	Pattie, James Ohio	The Personal Narrative of James O. Pattie	1833	American
1830–(?)	↓		de Fossey, Mathieu	Le Mexique	1857	French
1830–2	↓		Koppe, Karl Wilhelm	Mexikanische Zustände aus den Jahren 1830 bis 1832	1837	German
1831–2	↓		Tudor, Henry	Narrative of a Tour in North America	1834	American
1832–3	↓	↑	Becher, Carl Christian	Mexico in den ereignissvollen Jahren 1832 und 1833	1834	German

Travel literature (continued)

Date of observation	Direction of travel	Name	Title	First publication	Nationality of author
1834	]	Latrobe, Charles Joseph	The Rambler in Mexico: 1834	1836	English
1824–34	[(from Tampico)	Burkhart, Joseph	Aufenthalt und Reisen in Mexico in den Jahren 1825 bis 1834	1836	German
1830–7	↑	Ferry, Gabriel (Louis de Bellemare)	Vagabond Life in Mexico	1856	French
1838–9	↓	Blanchard, Pharamond	San Juan de Ulua	1839	French
1839–39	↑	Maissin, Eugène	The French in Mexico and Texas (1838–1839)	1839	French
1839	↓	Calderón de la Barca, Frances	Life in Mexico	1843	Scottish/Spanish
1840 (?)	Veracruz only	Coggeshall, George	Thirty-Six Voyages	1858	American
1841	↓	Mayer, Brantz	Mexico as It Was and as It Is	1844	American
?	↓	Norman, B.M.	Rambles by Land and Water	1845	American
1842	↑	Thompson, Waddy	Recollections of Mexico	1846	American
1843	↑	Payno, Manuel	in Viajes en México: Crónicas Mexicanas (1964)	?	Mexican
1843	↓	Gilliam, Albert M.	Travels in Mexico	1847	American
1845	↓	Heller, Carl B.	Reisen in Mexiko in den Jahren 1845–1848	1853	Austrian
1846	↑	Ruxton, George F.	Adventures in Mexico and the Rocky Mountains	1855	English
1846–65	Veracruz, various itineraries	Biart, Lucien	La Tierra Templada (1959) La Tierra Caliente (1962)	?	French

Travel literature (continued)

Date of observation	Direction of travel W	Direction of travel E	Name	Title	First publication	Nationality of author
1848-9		↓	Mason, R.H.	Pictures of Life in Mexico	1852	English
1848-51		↓	Robertson, William Parish	A Visit to Mexico	1853	English
1849-50	↑		Taylor, Bayard	Eldorado	1850	American
1852 (?)		↓	Ampère, Jean-Jacques	Promenade en Amérique	1856	French
1852-3		↓	Wilson, Robert A.	Mexico and Its Religion	1855	American
1851 ±	Veracruz & environs		Valois, Alfred de	Mexique, Havene et Guatemala	1861	French
1854-5	↑		Vigneaux, Ernest	Souvenirs d'un prisonnier de guerre au Mexique, 1854–1855	1863	French
1856	↓	↑	Tylor, Edward B.	Anahuac	1861	English
1856-7		↓	Müller, Baron J.W. von	Reisen in den Vereinigte Staaten, Canada und Mexico	1864	German
1864		↓	Kollonitz, Gräfen Paula	Eine Reise nach Mexico im Jahre 1864	1867	Austrian
1864		↓	Bullock, William H. (William Henry Bullock Hall)	Across Mexico in 1864-65	1866	English
1866		↓	Domenech, Manuel	México tal qual es: 1866	1922	French

Travel literature (continued)

Date of observation	Direction of travel	Name	Title	First publication	Nationality of author
1866 (?)	[	Elton, J.F.	With the French in Mexico	1867	English
186(?)	[	Altamirano, Ignacio Manuel	in Viajes en México: Crónicas Mexicanas	1964	Mexican

American military memoirs relating to the 1847 campaigns of the Mexican War

Name	Title	Date of publication
Beauregard, P.G.T.	in Williams, With Beauregard in Mexico	1956
Billings, Eliza Selu	The Female Volunteer, or the Life and Wonderful Adventures of Miss Eliza Allen, a young lady of Eastport, Maine	1851
Brackett, Albert Gallatin	General Lane's Brigade in Central Mexico	1854
Grant, U.S.	Personal Memoirs of U.S. Grant	1883
Jamieson, Milton	Journal and Notes of a Campaign in Mexico	1849
Kenly, John Reese	Memoirs of a Maryland Volunteer	1873
McWhiney, Grady, and Sue McWhiney	To Mexico with Taylor and Scott: 1845-1847	1969
Maury, Dabney Herndon	Recollections of a Virginian in the Mexican, Indian and Civil Wars	1894
Peck, John James	The Sign of the Eagle (Richard F. Pourade, ed.)	1970
Scott, Winfield	Memoirs of Lieutenant-General Scott	1864
Smith, George Winston, and Charles Judah, eds.	Chronicles of the Gringos	1968
Wilcox, Cadmus M.	History of the Mexican War	1892

Name

Title

Date of publication

Systematic sources

Period of observation	Name	Title	Date of Publication of Editions Used	Nationality
1803–04	Humboldt, Alexander von	*Political Essay on the Kingdom of New Spain*	Various editions	German
1830–32	Koppe, Carl Wilhelm	*Mexikanische Zustände aus den Jahren 1830 bis 1832*	1837	German
1836–43	Mühlenpfordt, Eduard	*Republik Mejico*	1844	German
1851–4	Richthofen, Emil K.H. von	*Die äusseren und inneren politischen Zustände der Republik Mexico*	1854	German
1830–50 (synthesis)	Sartorius, Carl	*Mexico about 1850*	1961	German
	Wappäus, J.E.	*Geographie und Statistik von Mexico und Centralamerika*	1863	German

reliability, and to set aside or at least to put into perspective imagination, exaggeration, and prejudice. They achieve a good deal in the way of context. By corroboration and other means they are able to make some judgments on reliability, but this is very difficult with material that is so highly subjective. They seem misguided with respect to the third objective, since that is where a great deal of the value of this literature really lies. Margo Glantz, in the valuable introduction to her *Viajes en Mexico: Crónicas Extranjeras* (1964), identified some of the major predispositions of nineteenth-century travellers to Mexico. The actual accounts are entirely absorbing, but the effect is diffuse.

A given volume of travel literature may be simply a sequence of journal entries published with cosmetic editing. Some authors set down recollections, assisted by journals and notes. There are compilations of letters written to a friend or a spouse, or material made to look like that. Some travellers sent serialized accounts to a periodical back home and then republished them en bloc. This is some of the best of what was purveyed in popular periodicals regarding Mexico. Inexpensive magazines, a fairly new phenomenon in western Europe, had gained tremendous circulation (von Mentz de Boege 1980). Their articles were generally of a pretty dismal quality, often derivative and bibliographically cavalier, indeed taken largely from the sources basic to this study. Illustrations were liberally redrawn; individual motifs from the better artists ghost through them without much attribution.

The typical traveller's account boils down to sequential narrative of movement interspersed with descriptions of the passing scene and reflections on a wide range of subjects. The basic structure is provided by the itinerary. Some of the most absorbing accounts detail not only the journey from Veracruz inland, but also a parallel inner journey, to use the formulation Graham Greene made memorable in his *Journey Without Maps* (1936): some overriding and organizing personal concern, which is another way of designating bias. The sensations generated in the one journey are often brought to bear on the other. Moreover, many travel accounts grade repeatedly from fact into fiction—at first a bothersome impediment to analysis. However, the "fiction" may be truer or at least more useful than the "facts," as Fussell has quite engagingly elaborated in his book on British travelling between the two world wars (1980:172–8).

The authors themselves may be grouped approximately according to background and motivation. There are the diplomats already referred to, as well as representatives of foreign companies, and a few merchants. Then there are those who had no business, so to speak, that is the naturalists, professional writers, and a heterogeneous remnant that can perhaps best be labelled as adventurers. One or two of these were prepared to call themselves just simply tourists. Between them they were able to marshal a good deal of imagery out of Scripture, classical mythology and Western literature. A few had scientific training. Most of them seem to have shared a propensity for the big gamble. They were prepared to assume the considerable risks involved in the journey in order to gain what each thought of as a potential fortune. Beyond all that, of course, there is a rich variety of temperament and many an original turn of phrase. The resulting landscape is impressionistic brushwork, full of highlights and shadows, a palette ranging from crude basic colours and a myriad of hues.

Diplomats

Some of the earliest observations available on independent Mexico appeared in *Notes on Mexico* by Joel Poinsett (1824). During the journey described he was "observing" on the instructions of President James Monroe; he was later appointed the first minister of the United States to Mexico. He was called a keen observer, "a professional appraiser of peoples and governments" (Gardiner 1959:14). Many Mexicans considered him officious, condescending, and a meddler in their politics—the North American attitude toward Latin America personified (e.g., Mares 1964:55). His journey and his commentary were hurried but diligent; he wrote in his journal while others took their siesta (1824:21). His private secretary, Edward Thornton Tayloe, left a diary that corroborates and expands upon the ambassador's material.

Brantz Mayer, a secretary of the u.s. legation in Mexico, made a journey in 1841, and Waddy Thompson, the "envoy extraordinary," made another a year later; both are said to have consolidated anti-Mexican feeling at home, and both make spicy reading. Their observations are strongly put and their images vivid. Both made their expansionism explicit and made invidious north/south comparisons. Profound animadversion runs through American commentary on Mexico in the early nineteenth century, diplomatic and otherwise.

H.G. Ward, English chargé d'affaires from 1825 to 1827, set pen to paper mainly to moderate strong fluctuations in enthusiasm among the British for investment in Mexico. His prose is circumspect and more reserved than that of the American diplomats, and yet condescension underlies it too. Besides investment advice, Ward's *Mexico in 1827* (1828) offers a systematic discussion of various aspects of the country, but this material is highly generalized and not of very much use in this analysis, except for the sections on foreign trade. The most relevant part is a travel account. The year was 1823 and Ward was still only one member of the commission that had come to prepare for official relations. He was in charge of transport. It was his responsibility to take the insufficiently rugged English carriages the party had brought with them through the coastal sands and up the much deteriorated Jalapa road—the road Humboldt had praised so highly twenty years earlier. Fortunately, Mrs. Ward, an accomplished artist, eventually came along this road as well. Her drawing of Jalapa is presented in Chapter 9.

Madame Calderón de la Barca, the Scottish-born and Boston-educated wife of the first Spanish ambassador to Mexico, may perhaps be grouped with the diplomats because of her privileged vantage point, and because, as one impressed reviewer of the spectacular new 1966 edition pointed out, she was identified with the mother country against which Mexicans had fought for freedom. However, her vantage point was curious on other grounds: "She was a woman in a country obsessed with *machismo* . . . an Anglo-Saxon and a Protestant in a nation that was part Latin, part Indian and overwhelmingly Catholic" (Johnson 1966:36). Her sharp-tongued, witty observations, her vivid language, and certainly the fame of her accounts set her apart from the other privileged observers, and indeed would distinguish her among the professional writers on our list as well. At some points one is inclined to agree with Manuel Rivera Cambas, Jalapa's dire historian, who felt the famous lady had been frivolous and had slandered those who had given her hospitality (1973,VIII:188–9). Her *Life in Mexico* (1843) is most useful here for its romantic interpretation of tropical nature, its renditions of exotica.

An account by Pharamond Blanchard, a member of the French punitive expedition from 1837 to 1839, approaches that of Madame Calderón de la Barca in several respects. The expedition involved a blockade of Veracruz throughout much of 1838, the bombardment and rapid capture of San Juan de Ulúa in November of that year, and an expedition ashore against Santa Anna in December. There were also long periods of waiting and various diplomatic forays by the commander of the operation, Rear Admiral Charles Baudin, until finally a treaty was agreed on in March 1839. Blanchard was in Baudin's retinue during much of the diplomatic travel, writing, drawing, and serving as a translator; he had spent twelve years in Spain. He was a romantic, sensitive soul, not always quite accurate in his terminology and frequently elliptical in his arguments, but often evocative.

Two accounts from members of the entourage of Maximilian and Carlota help us sense the changes in the journey upslope with the coming of the railway, but they may still be included in our company of authors since they add substantially to the characterization of the lowlands, particularly the terrain just inland from Veracruz, as well as the elaboration of a particular perspective on the lowland tropics. An Austrian countess, Paula Kollonitz, accompanied the imperial pair to Mexico. They arrived at Veracruz in 1864, when the new railway extended a bare fifty kilometres from the port, but already funnelled most traffic away from the Jalapa Road and in the direction of Orizaba. The duchess seemed to be looking at the country and its people through a lorgnette, noting the surroundings in terms of clichés: miasmatic swamps, desolate vegetation, picturesque natives. It had been her duty, she explained in the preface, to be polite and not to probe too deeply.

The emperor's press secretary, Manuel Domenech, a Frenchman, wrote a much less inhibited and very witty account of travel along the same route two years later. He described the rigours of travel better than any of the other authors: Veracruz was hot as a baker's oven, unconscionable prices were being charged everywhere, the rough roads "milled the bones."

Representatives of European Investment and Trading Companies

Like the diplomats, these travellers had excellent contacts, were given privileged information, and were backed by policy decisions. The agents' accounts are generally positive in tone. They needed to present Mexico in a favourable light in order to maintain the interest of the stockholders.

The Rheinish-Westindische Kompanie, which pioneered German trade in Mexico but was in disarray by 1830, sent a sub-director, C.C. Becher, to review the situation at first hand (Pferdekamp 1958:49). He came with a letter of introduction from Humboldt, the supreme recommendation. He also carried Humboldt's appreciation of the tropical environment and reservations about the people living there. Becher has excellent material on both Veracruz and Jalapa.

William Parish Robertson arrived in Veracruz from England in 1851 on behalf of the Committee of Spanish American Bondholders, with the responsibility of renegotiating a loan. He wrote optimistically on prospects for British capital, especially in mining and railway building as well as on immigration from the mid-latitudes. His account is extremely discursive; its two volumes were published for the author. He is often pompous, and at times sentimental. He seems always to have been concerned to note the fine class of his contacts. When he comes to his description of Jalapa, he quotes what a number of other authors had written about the town and its environs, a kind of anthology of paradise. He felt his readers might otherwise regard his own description of this favoured region as excessive.

It is not clear just why Alfred de Valois, a French traveller, came to Mexico, but he wrote most interestingly of the world of commerce and was well connected. The French consul in Veracruz personally escorted him on an excursion outside the walls and took him to various social events.

Merchants

William Bullock, who was something of a naturalist, was made to reflect on his motives for travel when he presented his letters of introduction to the principal merchants of Veracruz: "As I brought no cargo nor consignment, and had not any speculations to offer, those to whom I presented them, after a few questions, generally left me with marks of surprise, that a man in his senses could venture so far from home to such a place, with motives so inadequate" (1825:1,25).

A laconic account of a more sensible journey was left by a sailor and petty merchant, George Coggeshall, who came to Veracruz some time between 1830 and 1840. He briefly described the port and noted that "generally strangers only visit it for the sake of gain" (1858:56). The sale of the goods he had brought with him from New York netted a profit of 150 per cent.

Some of the hardships associated with commerce in the Mexico of the 1820s have been preserved in the story of James Pattie. He arrived in Veracruz after six years of journeying westward and southward overland from St. Louis. This was to have been the one big fortune-making adventure of his lifetime. When Pattie sailed for home from Veracruz, he had long since been robbed of the goods he had brought to sell. His claims for restitution to various government offices in Mexico City had been unsuccessful. Finally he was reduced to the charity of sympathetic listeners for the fare home. The account of his suffering was widely read in the United States; it exemplified Mexican lawlessness and helped to fuel the resentment that eventually led to war.

Even though the commentary left by the merchants themselves is scanty, they are a constant presence in the literature basic to this analysis. Travellers sometimes heard their loud profanities at the tables under the Veracruzan arcades; or saw how they disported themselves at the riverside resort of Medellín. Valois noted with considerable disdain that the Mexicans could not forgive the foreigner who by dint of courage, patience and intelligence, had made a fortune. Lazy and jealous, they did not esteem active men (1861:90). Mayer described the entrepreneurs as those "who have come to die of the *vómito*—or, to make their fortunes (if they survive the certain attack of that disease) and return with shattered constitutions to colder climates, to ache in memory of the heat they endured in Mammon's service" (Mayer 1844:3).

The Naturalists

They came without official charges or commercial interests. They were not entirely disinterested; one was also a collector, another acquired a mine, still another became a *hacendado*. But they all had a more than casual fascination with nature, particularly the world of plants, and a determination to gather information at first hand. They were not content with a view from the stagecoach window but travelled by horseback with only a few companions and thus were free to leave the main roads. And they were all German—whatever that may mean.

Carl Heller and Baron von Müller epitomize these tendencies among our group of travellers. Heller, only twenty-one in 1845 when he began the journey, had been trained in the natural sciences and was to become a respected botanist (Pferdekamp 1958:222). However, his purview was wider. He hoped to contribute to the debate over Mexico as a destination for German emigrants. He is most interesting when he describes the emotions stimulated by the landscapes through which he passed.

Müller came to Mexico in 1856, some eight years after Heller had completed his travels, and indeed after most of the other authors considered in this study had also come and gone. He had aspirations of a Humboldtian scale; he wanted to write a natural history of Mexico that would supersede anything done in the intervening fifty years. He came with a doctorate, belonged to several learned societies, and had excellent connections in Mexico. He might well have made a major contribution, but was prevented from even beginning the envisaged treatise by the loss of the bulk of his notes and specimens in transit—the ultimate nightmare of any academic. He was reduced to publishing a travel account based on the diary he had carried with him, and such materials as he was able to obtain later from Mexico by mail.

In the accounts of Müller and Heller, the results of scientific investigation are intermingled with romantic passages. Müller described the anatomy of the firefly along with the engaging way in which the ladies of Veracruz used captive fireflies as live jewellery during their evening *paseos*. On an excursion into the high tropical forest on the lower slopes of the escarpment near Córdoba, he was enraptured by the luxuriance of his surroundings. Germanic myths came to him while resting on the banks of a river. In the next sentence he tells how he went down to measure the temperature of the water.

Müller had a full complement of antipathies. He was anti-Catholic and quite caustic too about loud-mouthed, overconfident North American travellers he encountered en route. He was condescending about Indians in a manner virtually standard throughout this literature, and highly disparaging of Mexicans with a mixed racial background. He seemed especially repelled by the typical lowlander, the "jarocho."

It is difficult to classify the English traveller Edward Burnett Tylor. He was a traveller of academic if not scientific or naturalistic bent and thus perhaps fits better here than with other travellers. He was a student of cultures, later to become known as the founder of cultural anthropology. Of special interest then, are his comments on the people of the lowlands, in particular those on Indian home life in the mid-slope country just south of Jalapa. He entered the country in March of 1856, the same year Müller came, and like him he left Veracruz on a carriage pulled along the newly built railway that extended some distance out of Veracruz in the direction of Córdoba. He thus helps us end this book's epoch: by the time he arrived in Jalapa it had been superseded in trade by Orizaba.

The well-known mid-century portrait of Mexico by Carl Sartorius deals in detail with the various habitats along the Jalapa road and also includes a very useful account of a journey through the wetlands of the San Juan basin just west of the port of Veracruz and south of La Antigua. It may be considered both a systematic source and an account of travel. His was not a compendium and he made no claims of comprehensiveness. Instead, he employed a lively style, and imaginatively interwove systematic description with exemplification. The book is also superbly illustrated by Moritz Rugendas. The English translation has been used in this analysis; a Spanish translation is in preparation in Mexico.

What Sartorius said of the country and particularly its tropical lowlands was based on a longer period of observation than that of any other of our authors. He came to Mexico in 1824 as an idealistic young man, a political fugitive, strongly for a free and united German nation (von Mentz de Boege et al. 1982:234). He stayed, except for a three-year return to Europe, until his death in 1872. His initial purpose was to establish a utopian community of German settlers, which was only partially successful. Sartorius did become a prominent *hacendado*, a producer of sugar cane, coffee, tobacco, corn and cattle, indeed a leading example of the successful nineteenth-century German entrepreneur abroad, and one of the very few to do well in Mexican agriculture (Koppe 1837:1,54–5). The central buildings of his property, El Mirador, may still be found just north of Huatusco. This hilltop site is something of a botanical garden to this day. It has been used as a base for field research by a succession of scholars.

Sartorius showed little dread of tropical nature; he travelled widely within the lowlands and reflected well on what he saw. He shared Humboldt's reservations about the lowlanders and did not approve of the way they lived, but he could at least treat the subject with some humour, and every so often one detects a wry respect for their ways of making do. He had committed himself to agricultural development and immigration in this region. It is a bias similar to that of corporate representatives concerned with the maintenance of investments. The geographer Ratzel thought Sartorius had been particularly energetic and just plain lucky; most emigrants would be neither (1878/1969:382). In fact, most schemes for German immigration into Mexico did misfire. The main streams of the movement eventually were directed to North America and Brazil.

Karl Wilhelm Koppe's work on Mexico in the early 1830s seems little known, but is in fact a highly valuable systematic source. Koppe also sent home travel accounts, many of which are collected in his *Briefe in die Heimat* (1835). Two such passages are appended to his systematic treatise, which has not yet been translated from German. They deal with his journey downslope and the wait for embarkation. He was detained in Medellín, a small village on the Jamapa River south of Veracruz. The port of Veracruz, under the control of Santa Anna, was in rebellion against the federal government of Bustamente, who had countered the insurrection with a siege. Koppe needed to bypass the town and reach his ship out in the anchorage. While he waited for news from a scout he had sent ahead, he explored the forested surroundings, as though this were an opportunity for a bit of amateur science too good to miss, and left a description rich with wonder.

Curiosity was the most engaging quality of the naturalists. They measured heights, corrected the figures of those who had measured them wrongly, enumerated rivers and lakes, and made lists of Latin names. The classification of plants concerned them particularly—a legacy of Linnaeus. However, they often noted interrelationships among plants as well, and knowingly or otherwise described plant communities. Links were made between plants and environmental factors, particularly altitude. However, such gathering of facts and the profession of scientific objectivity was something of a fashion of the time; the Latin binomial was a kind of ornament.

It is often apparent that the naturalists visiting Mexico were interested in natural rather than biblical explanations of the phenomena they encountered and had intimations at least of the radical new ideas of the day about natural history. Often, however, in their "scientific" forays, they seemed to be practising mechanistically what they did not well understand. And they were not yet completely secularized either. Loren Eiseley has noted in many naturalists of the nineteenth century a "traditional Christianity overlaid by a wash of German romantic philosophy. Elements of the new science and the new discoveries [were] being fitted into what [was] regarded as the foreordained design of the Creator" (1961:95).

In a grove of royal palms that seemed a natural cathedral, Sartorius "bowed before the All-Wise" (1858/1961:7). Müller, the most relentless horseback traveller through the lowland tropics among our entire complement, profuse in his description of plants and animals, as well as biting in his comments on incompetence and superstition, felt compelled to consent in his preface to a more than slightly superstitious saying: "None who wander under palms remain unpunished" (1864:1, vii; translation by the author). He had lost his trunks and his health, reason enough to ruminate about punishment, but there is also an echo here of deeply rooted northern unease about southern places: an irresistible attraction shaded by warning of what might happen there to body and soul.

Professional Writers

Only a few of the travel accounts may be considered travel literature per se, that is, the work of professional writers en route responding to the passing scene, finding in it the stimuli for their art and points of departure for the expression of their varied predispositions. Some were under contract, others no doubt just hoping that eventually there would be recognition and a market.

Several were journalists of considerable talent. Bayard Taylor became well known for the material he sent to the New York *Tribune* on gold rush California. He returned home from the West Coast via Mexico, the Jalapa road, and Veracruz. An unnamed war correspondent from the *Boston Advertiser* accompanied the American army in 1847. His description of Jalapa is densely packed with information not available elsewhere (Smith and Judah 1968:217–22). Thümmel's anthology of journalistic Mexican travel sketches includes a very effective but also incompletely attributed piece on the view of Veracruz from the deck of a ship anchored between the fort and the town (1848:89–109). Various authors whose work on Mexico fits primarily into other categories, such as Koppe and Becher, also wrote for periodicals.

Gabriel Ferry, whose real name was Louis de Bellemare, and Lucien Biart have provided us with mixtures of fact and fiction. Elaborate dialogue immediately suggests fiction; the biographers of both authors confirm that it was invented (Prévost and d'Amat 1951, v:1341; Biart 1959:271–5). Both authors spent years travelling the byways of Mexico, including intricate itineraries through the lowlands of Central Veracruz. There are many indications in their prose that they had observed environment and culture carefully. Both contributed to the delineation of a subculture, that of the *jarocho*, which will be discussed in Chapter 7. When these authors take us to horse races, *fandangos*, or the wrecking of a cargo ship on the shore at Boca de Río during a *norte*, we are given information which may not be as clearly specific as the "data" in the literal accounts, but which is not likely to have been much more distorted or more imaginatively rendered.

Fact certainly shaded into fiction in the novels of Charles Sealsfield, a German author, whose real name was Karl Postl. He dealt with various Mexican subjects and introduced one of them with an excerpt from his journal, in which he describes his approach in 1828 to the coast of Veracruz and the beginning of the journey into the interior (Sealsfield 1974). There is some controversy over whether or not Sealsfield ever was in Mexico at all; there are certainly some elliptical locational references in the excerpt from his alleged diary. Of interest here is Sealsfield's deliberate redirection of North European yearnings southward, from the Mediterranean to Mexico, and his repeated allusions to the ambivalent feelings its landscapes and post-Independence conditions evoked. If all this was derived from other authors it is at least a reinforcement of some important basic aspects of the foreign visitor's view of Mexico and in particular its tropical lowlands.

Others

A good number of the travellers—Latrobe, Gilliam, Vigneaux, Tudor and William Henry Bullock Hall—fit none of the above categories. They seem by turns to be dilettantes, fugitives, adventurers, or just simply tourists. The most interesting are agitated by some private concern. While they cover the kilometres, they are also on an inner journey. Robert Wilson, a lawyer, was thus compelled. He was determined to use the laws of evidence to debunk Prescott's account of the Conquest of Mexico. Fortunately, he also gathered bandit lore and retold old scandals about the monastery of San Francisco in Jalapa. Various of the casual travellers were fundamentally disgusted with Mexico. They had brought this feeling with them and it deepened as they went along. George Ruxton, an English adventurer, for example, denigrated the country and its people more blatantly than even the two diplomats from the United States, Brantz Mayer and Waddy Thompson, but not yet in quite as sophisticated a manner as the geographer Friedrich Ratzel, of whom more will be said. Ruxton maintained in his preface that "if the Mexican possesses one single virtue, as I hope he does, he must keep it so closely hidden in some secret fold of his *sarape* as to have escaped my humble sight, although I travelled through his country with eyes wide open, and for conviction ripe and ready" (1855:iv).

J.C. Beltrami, an Italian liberal and romantic, was convinced that Mexico was indeed uncivilized, but he nevertheless expressed the hope that he would be able to redress some of the unmerited evil spoken of the country. Jean-Jacques Antoine Ampère, a French traveller, whom his editor calls "un simple turista," but who occasionally took accurate aim, compared those who were finding Mexicans and Mexico contempt-ible to the famous vultures of Veracruz, preoccupied with carrion (in Glantz 1964:415).

One of the most absorbing authors in the entire theatre troupe is Mathieu de Fossey. He came to Mexico from France in 1830 with a group of peasants who had long worked his family's lands. They were to participate in the ill-fated Coatzacoalcos colonization scheme. After relating something of that dismal experience, he then described his travel northward to Veracruz, westward through the lowlands, and sub-sequently along many other Mexican roads over a period of several decades. He was in search of employment, it seems, and an agreeable place to settle down. He had been to university, knew Latin and Greek, but did not have sufficient English or Spanish, nor the accounting skills necessary to find a place in a commercial establishment. His education simply had not equipped him for his venture into the New World. Yet he was a careful observer and he wrote with imagination. He did not hesitate to speak of his dreams and seems to have escaped embitter-ment.

There are whole shelves of volumes of more casual travel literature. They seem too solidly bound, too long preserved. Two such accounts, the books by B.M. Norman and R.H. Mason, were selected as examples. They are as rich in clichés and as muddled in their facts as one could want. They were consulted at each stage of the journey in order to maintain an appreciation of the better material.

Other, more readable "controls" were considered as well. Various Mexican historians provide alternative realities, particularly with respect to the difficult living conditions in the towns and countryside. Foreign observers missed much of that. Then there were the accounts of several Mexican travel writers and a spate of Veracruzana from the Editorial Citlaltepetl.

The first of the Mexican writers used is Manuel Payno, a young romantic with good political connections. He was to climb through a long series of governmental posts and to become a writer of some accomplishment. The second is Ignacio Manuel Altamirano, highly esteemed in Mexican letters. Both knew very well in what negative tones foreign visitors spoke of Mexico and were only too well aware of their country's disorder. Both were also staunch patriots; they are considered here in order to see what difference that made. Both were uplanders and might be expected to find the lowlands almost as strange as would the foreigners. Payno, at least, was quite as hapless and peremptory in his explanation of lowland conditions as one of them might be. Both also left some fine prose. Altamirano, for example, included an excellent passage on the uplanders' fear of the "zona mortífera," that part of the lowlands where yellow fever could be contracted. Both authors were able to make light of many of the rigours of travel that often soured the foreigners, and they could outdo the best of them in romantic rhapsodies.

As for the Editorial Citlaltepetl, it was for many years a source of what a modern Mexican in a colloquial moment might call super-Veracruzana. It was headed by Leonardo Pasquel; he issued or reissued numerous local histories, notably the voluminous *Historia de Xalapa* (1959) by Manuel Rivera Cambas, as well as various colonial Relaciones Geográficas and other important documents for the study of the lowlands. He put out bibliographies, biographies of famous sons and daughters, poetry—all furnished with uniformly laudatory introductions by himself. Unfortunately the material is often incompletely attributed and poorly produced. It has been consulted here for its enriching lore.

The prime concern remains an assessment of outsiders' views; we need to take some note of how these views were received in North America and Europe. It does seem, in fact, that the travel literature was widely diffused and that it influenced popular thought at various levels. Even if one takes prefaces with a grain or two of salt, those in this literature often do indicate pressure from publishers or businessmen or friends to get quickly into print. Some books went into several editions. Various of the systematic sources were translated. Thoughtful analyses of this literature, such as those of Pferdekamp (1958) and Ortega y Medina (1955), to which repeated reference will be made, explicitly and implicitly indicate a wide diffusion. Practical information on Mexico was needed in various quarters. Investors wanted to assess the possibilities; there is evidence that in England, at least, investor enthusiasm waxed and waned according to what travellers reported. Many of the landless, especially on the Continent, were considering immigration; most were dissuaded by what they read, especially the travel articles published in the popular magazines. Other Europeans found vicarious adventure in this literature. The judgment must remain qualitative, but it is apparent that Mexico was selling well in the bookshops.

Military Memoirs

The recollections of the men of the American army, in 1847 moving from the coast into the interior for the concluding phases of what is commonly known as "The Mexican War," include some useful commentary on several aspects of the lowland tropical landscape; the soldiers also repeatedly characterized the Mexican people. The list of military authors in Table 1 includes, besides the memoir of the commander, a good number of the accounts relied upon by various historians of the Mexican War. The account of Private George Ballentine is prominent among them. He "wrote what is perhaps the best account of Scott's campaign by either an enlisted man or an officer." This is a huge literature; the sampling undertaken here was facilitated by two useful recent bibliographies (Smith 1969; Gunn 1974) and two anthologies (Smith and Judah 1968; McWhiney and McWhiney 1969).

The soldiers' accounts, both published and unpublished, spread information widely throughout North America and contributed to the low esteem in which the Mexican was held. Common priorities and antipathies pervade these accounts. Most of what was recorded dealt with camp life, the logistics of war and the actual fighting. This was a world precisely ranked; in it, intrigue was endemic and promotion the highest good. Under such circumstances, it was often considered necessary to set the record straight. Embittered First Lieutenant P.G.T. Beauregard provides an extreme example. He had not received the credit he thought due and offered a pathetic explanation of his unsung achievements during the siege of Veracruz and the battle of Cerro Gordo. The man who received much of the credit Beauregard felt he deserved was another lieutenant, Robert E. Lee (Williams 1956). The Mexican War provided an excellent start for many other officers who were to distinguish themselves in the American Civil War.

What the soldiers noticed of their surroundings was usually phrased in cryptic, conventional terms. Most did not have the background for more; very little of Humboldt, for example, is traceable in this literature. A few recognized this and wished it were otherwise. When the soldiers did write of the heat and the thorny, scrub-like vegetation between Veracruz and Jalapa through which they pursued an enemy that was not fighting according to the rules of civilized warfare, they often became eloquent. It is mainly for this that the accounts are included in this analysis.

Systematic sources

The travel accounts and military memoirs can be set into a good factual context. Seven authors compiled and in various ways synthesized information on Mexico at intervals throughout three-quarters of the nineteenth century: Humboldt, Koppe, Mühlenpfordt, Sartorius, Richthofen, Wappäus and Ratzel. Their books helped to prepare the more thoughtful travellers; they were intended and indeed taken as guides to statecraft. As it happens, the nature of most of this material is attributable to developments in academic geography within German universities. The books, in turn, became basic texts. All in all, therefore, their influence on conceptualization regarding Mexico and the tropics will have been considerable.

The early nineteenth century has been called the classical period of geography and the names most prominently associated with it are Alexander von Humboldt and Carl Ritter (Hartshorne 1939:48–84). Both were concerned with objective observation; each accumulated vast amounts of field data and each attempted synthesis, but with somewhat different criteria. Humboldt broke new ground in *systematic* geographic studies, elaborating basic principles on the geography of plants, climates and landforms. Ritter drew his materials into relationship in *regional* terms, that is, by cultural or political realms. Unfortunately, he himself did not deal in detail with the Americas. In subsequent elaboration this approach proved something of an intellectual cul-de-sac in that it led to rather arid itemizing and categorization.

Alexander von Humboldt came to Mexico at a fortunate moment. Mexico was becoming increasingly receptive to new ideas; its high society was already extensively affected by European, and especially French, thought and practice. When Chappe d'Auteroche made his journey to California in 1769, for example, to observe the eclipse of the sun, he had been accompanied by guides and his movements had still been severely restricted (Chappe d'Auteroche 1778). By the turn of the century such official suspicion of the discerning foreigner had diminished a great deal. Moreover, enlightened monarchs and their chief administrators had perceived a need for information of many kinds and were facilitating its collection. The Crown, during the administration of Viceroy Revillagigedo (1790–94), had ordered a census of New Spain. Humboldt, who carried excellent recommendations from the Spanish court, was given access to this and a great deal more. Receptivity and new data thus came into fortunate conjunction with the talents and ambition of an already prestigious German scientist.

Humboldt showed sympathy for the move toward Independence and made analyses which could be used by the new administrations. The Mexicans respected and indeed sometimes adulated him to the point of absurdity (Miranda 1962:106–7, 205). His first maps and compendia influenced foreign investment and statecraft vis-à-vis Mexico. The *Political Essay on the Kingdom of New Spain* (1811) elaborated on these and became required reading for anyone going to the new country. One could, with luck, expect to see what Humboldt had seen; one might even be able to add to or update his factual information. To the thoughtful as well as to the more pragmatic reader, the *Essay* was a model of disinterested analysis: this was *science*.

For our purposes it is diagnostic of a particular approach to the tropics. In places it may still have practical use. A modern pleasure sailor, for example, could do worse than to rely on Humboldt's description of the weather of the Gulf.

The *Essay* has its limitations (Miranda 1962; Ortega y Medina 1960). Humboldt spent only one year in Mexico, hardly enough time even for someone as diligent as he to do both the inviting archival work in the capital and field exploration of the kind he had done in South America. His passage through the coastal lowlands, in particular, seems to have been hurried. Humboldt relied heavily on the information being assembled by Mexican scholars and governmental functionaries during the reforming, information-hungry vice-regal regime of Revillagigedo—sometimes without adequate citation, it seems. He may also have been compromised by the very connections that had given him such excellent access to official data in the first place. His elite hosts' biases regarding the indolent native, for example, seem to have become or at least coincided with his own. It is likely, moreover, that he had to restrain his criticism of the conditions he found. One can find other, lesser, faults. Much of Humboldt's prose is disorderly and prolix; his thought was often not very coherent. All that, however, is overshadowed by the merits of the *Essay*: "Seen in the context of its time, its true value lay in its modern and variable focus, in its sweeping visions and intuitions, its penetrating glimpses of the future, in the causal sequences and factual interrelationships it proposed, as well as its many generous, elevated thoughts" (Miranda 1962:173; translation by the author).

Karl Wilhelm Koppe's little known *Mexikanische Zustände aus den Jahren 1830 bis 1832,* untranslated except for some excerpts offered by the Editorial Citlaltepetl, provides a richly detailed, engaging portrait of Mexico on the morning after Independence. Koppe had read widely in preparation for his journey. He came as the first consul of Prussia, yet he sampled at more prosaic levels than Humboldt and was obviously freer to criticize. He was particularly well versed in economic matters, but broadly curious and sharply observant. He seems to have mastered Spanish and involved himself considerably in New World literature and art (Pferdekamp 1958:201–2).

Koppe's *Zustände* are largely *Misstände*, which is to say that his book is mainly an indictment, but it was phrased with care and some sympathy. His two-volume work contains several different sorts of material. There is an introductory history of the revolutionary events from 1810 to 1832, then a factual description of each of the states of the republic. It is evident that Koppe searched carefully for official documents; his citations are explicit: others of our authors were often cavalier in this respect. Before he published his systematic study of Mexico, Koppe had attained a considerable readership with his travel accounts, published as *Briefe in die Heimat* (1835). In *Zustände* he drew on this to discuss what it was like to travel in Mexico and then, in conclusion, described his departing journey toward Veracruz, where he never arrived (he had to leave from Alvarado). His *Zustände* amounts to an evenhanded and sensible assessment of the prospects for the investor, the trader, and the emigrant. The overriding concern of the book is the lawlessness on every hand.

In 1844 Eduard Mühlenpfordt published a compendium of what he claimed was worth knowing about Mexico at the time. The book is based on a seven-year residence in Mexico during which Mühlenpfordt served for some time as a director of road construction in Oaxaca. It was evidently meant as both a practical guide and an academic sourcebook. It is apparent that Mühlenpfordt derived much of the material relevant to the Jalapa road either from Koppe or at least from the very same documents used by Koppe—unfortunately, without acknowledgment. Mühlenpfordt did update some of Koppe's material, as there was a seven-year gap between their books, but in various key areas, such as foreign trade, the figures in the two accounts are the same.

Mühlenpfordt emigrated in a group that included Carl Sartorius. Both of their books have been republished recently in English. Sartorius is definitely the more readable of the two authors, but Mühlenpfordt is more factual and precise. Both were romantics: Sartorius allowed this tendency a freer rein while somehow avoiding the profound animadversion which permeates Mühlenpfordt's work.

The punctilious, densely written, and often eloquent book on which we have already relied for a survey of Mexico's foreign relations, Emil Karl Heinrich von Richthofen's *Die äusseren und innerin politische Zustände der Republik Mexico*, was published in 1854. It is, unfortunately, as little known as Koppe's *Zustände* and like it has never been translated. Richthofen was the representative of the Prussian court in Mexico from 1851 to 1854. His view was from the capital; the book is organized under the headings of government ministries and ranges over all major aspects of the life of the nation. It was clearly designed for the arriving diplomat or the policymaker at home.

From a paragraph near the beginning of the book we may sample his care and restraint, as well as his perspective on the difficult post-independence period, the *época negra*. He calculated that during the thirty-three years of the republic's existence there had been at least ninety different people in presidential and ministerial positions, each with his own complement of political and administrative ideas. Then he observed that:

> Anyone who has carefully followed Mexico's governmental history must admit that the higher officials have, on the whole, had a working knowledge of political and administrative procedures and that almost all have had a comprehensive understanding of what was in the best interests of the republic. However, because of a lack of strength, a failure to give stability to the country's constitutions, and the continuous revolutionary movements thus facilitated, good ideas and sensible measures could never take root. (Richthofen 1854:6–7; translation by the author)

Richthofen hoped that Santa Anna would yet turn out to be the strongman Mexico needed, and believed that some day the country would achieve international prominence. This was to happen over a century later, when it became a leading oil producer—and then a leading Third World debtor.

A regional geographer, Johann Eduard Wappäus, provides the factual endpiece for our analysis (*Geographie und Statistik von Mexiko und Centralamerika* 1863). He was never in Mexico, but travelled elsewhere in the American tropics. His book is a synthesis of virtually all the systematic studies used here, from Humboldt onward, as well as many of the travel accounts.

He included a comprehensive bibliography, which was a useful checklist. Wappäus, a professor of geography at Gottingen, is said to have affirmed that geographers needed to concentrate on the gathering of facts with no presuppositions (*Allgemeine Deutsche Biographie*, 1896, Vol. 41:164). He was following Carl Ritter and using all the arbitrary categories common to the specialty. He began with the physical geographical basics, continued on to population, then to products, production systems and trade, and concluded with things of the spirit: the church and statecraft. Unfortunately, there is also more than a touch of the pedantry to which the practitioners of regional geography have often been inclined.

It is difficult to avoid distending the chronological limits somewhat at this point and taking the geographic succession a step further to include Friedrich Ratzel, who travelled in Mexico in 1874 and 1875. He wrote at length and with strong language on tropical nature in Mexico's lowlands; he made explicit references to some of the descriptions of it in the travel literature under analysis here, taking issue with various points made and re-expressing others in dramatic terms. We meet him as a young man writing travel sketches for the *Kölnische Zeitung*. Eventually he was to become famous for his systematic study of human geography, especially for his attempt to give it a scientific basis. He is considered an heir of Humboldt and it is interesting to see the direction in which Humboldt's ideas had grown.

In his preface, Ratzel denied having any literary or scholarly pretensions, but affirmed that this did not mean he had relaxed his habitual discipline to avoid hearsay and superficiality. He was reporting without preconceptions what he had seen. What he actually produced was a relentless critique of all things Mexican, tropical or otherwise. He dismissed the country's struggle for independence, subsequent political movements, intellectual life, and high society. He found its people corrupt, incompetent, and unbelievably indolent. He thus outdid even the most strident of the North American visitors on our list, setting up a model of disparagement.

PHYSICAL CHARACTERISTICS OF THE VERACRUZAN LOWLANDS

The terrain over which the travellers constructed a fanciful landscape— monumental in some places, hideous in others, but always rich in nuance and thoroughly intriguing—can actually be quite dreary, especially during the dry season. These are the months from late fall to early summer, which was when most of the visitors came. The scrubby vegetation is dormant and seems quite dead; only the wetlands and the water courses are green. The topography along the road is not dramatic, except at the major river crossings and at a few points when one can look down into canyons. Temperature and humidity are usually high. Even now, with good roads and sizeable settlements along the way and many opportunities to take refreshments, it tends to be country through which one passes as quickly as possible.

It is useful to sketch it out—in two dimensions—in order to provide context for the specific references in the individual chapters. Also, since the shifts in the perception of the landscape and even the inhabitants are often cued environmentally, it is particularly important to fix the location of certain key features in advance.

The volcanic peak of Mount Orizaba, some 140 kilometres directly west and inland of Veracruz, was the first feature of the mainland that the travellers arriving by sea could expect to glimpse from the deck of their ship. Soon thereafter they would become aware of the low shoreline of dunes around the port (Figure 1). Looking down on these dunes (Figure 2), one notices crescent shapes: the tips always point downwind. In this case they reflect the dominant direction of winds during the winter storms, the *nortes*. The youngest dunes are still alive, older ones

2 Dunes of various generations, driven by *nortes*; swamps, formed ahead of the live dunes advancing over the lobes of earlier dunes, considered to be the source of the vapours that carried yellow fever

have been fixed by vegetation. The whole belt is pockmarked with depressions in which moisture collects and sustains hydrophytic vegetation.

The dunes are bounded on the landward side by a belt of humid lowlands, including many wetlands (Figure 1). This belt is presented by nineteenth-century travellers as fairly continuously forested, with enclaves of hydrophytic vegetation or open water in the wetlands proper. It has since been extensively cleared and drained: the main use is ranching.

From the humid lowlands the terrain slopes upward toward Jalapa. It has been deeply incised by a series of eastward-flowing rivers. The thick stratum of conglomerate rock just below the surface commonly weathers into cliffs, resulting in some spectacular canyons known in Mexico as *barrancas* (Figure 3). A dryer climate and the remains of a low forest prevail over this dramatic topography. It has always been laborious to travel across the grain of this landscape. One of our travellers found that a narrow barranca over which one could easily converse took two or three hours to cross (Tylor 1861:31).

3 A view northward over the incised sloping plain east of Jalapa

At about 1,000 metres above sea level the travellers usually sensed a slight drop in temperature and noticed scattered oaks, which became a symbol of relief. Humboldt crossed the lowlands in 1804, at a time when a new road between Veracruz and Perote was under construction, the same road that travellers coming a few decades later would see in ruins. He was informed that the builders intended to erect stone markers at intervals indicating not only distance, but also elevation above sea level (Humboldt 1812, Band 1:288). They would assure the ascending traveller that relief was approaching.

Average precipitation figures increase at about 1,000 metres, sustaining a more luxuriant vegetation (Figure 4). Tropical and temperate species are intermingled in both the forests and the cultivated fields. Jalapa, at about 1,500 metres, lies within this favoured zone. Above Jalapa, average temperatures drop further and mixed stands of conifers and oaks set in, often shrouded by mists (Figure 5). Pure stands of pines surround the pass at about 2,400 metres; beyond it the landscape opens abruptly on to a vast, much dryer plateau.

The seasonal rhythm over the lower levels of the coastal landscape is a fluctuation between the wet and the dry. Frequent frontal rainstorms occur between May and October, and a characteristic daily sequence

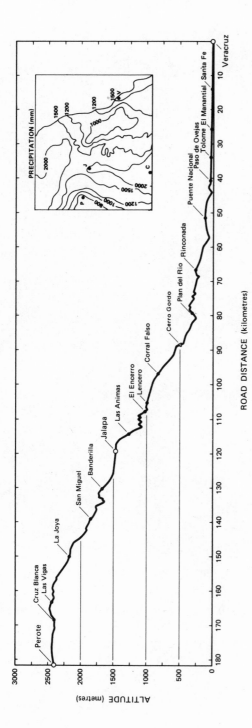

ALTITUDE (metres)

ROAD DISTANCE (kilometres)

PRECIPITATION (mm)

Perote
Cruz Blanca
Las Vigas
La Joya
San Miguel
Banderilla
Jalapa
Las Animas
El Encerro
Lencero
Corral Falso
Cerro Gordo
Plan del Rio
Rinconada
Puente Nacional
Paso de Ovejas
Tolome
El Manantial
Santa Fe
Veracruz

4 Altitudinal profile from Veracruz to Perote and the average annual precipitation in the region

5 Stands of conifers and oaks above Jalapa

obtains: fine mornings, clouding around noon and rain soon thereafter. During the nineteenth century this was the time of maximum danger from yellow fever. Late in the wet season there is always the possibility of hurricanes. The dry season, roughly from November to April or May, which coincides largely with the northern hemisphere winter, is the season of *nortes*, the southward extension of weather systems generated over the North American continent. They are interspersed with bright and calmer weather, the optimal "windows" for the approach of sailing ships to the Veracruzan shore. The danger of disease in the one season and shipwreck in the other, plus all the additional problems of life in the lowlands, were considered by more than one observer as the colony's and the new nation's best defences.

Around Jalapa, two broad climatic patterns interdigitate: "wet and dry" with "summer and winter" or simply "temperate" with "tropical." This leads to an incredible environmental complexity as will be seen in Chapter 9. Perhaps three-quarters of the trees drop their leaves in December and January, and some even assume autumn colours before they do so, which is always much appreciated by visitors from the mid-latitudes. Mist and rain can come to Jalapa and points upslope throughout the year, but they are particularly frequent during the rainy season, and particularly miserable near the summit during the winter.

The altitudinal stratigraphy of this landscape obtruded on the observations of the nineteenth-century traveller. Four major levels were usually distinguished: the hot and humid lowlands proper below 50 metres above sea level; the hot and dryer, often dismal slopes between 50 and 100 metres; the narrow band of amelioration around Jalapa from 1,000 to 1,500 metres, for those moving upslope as well as those going in the other direction; then the colder air and increasingly coniferous forests growing over a volcanic litter just below the pass.

PERIOD OF OBSERVATION

New Spain's exposure to foreign scrutiny and influence increased dramatically with Independence, but the increase was already foreshadowed in the changing nature of its foreign trade. There had long been a lively trade in contraband, of course. The official liberalization of legitimate trade had already begun in 1778 with the lifting of the monopoly of Cádiz and the authorization of trade with Mexico out of most peninsular ports, however, it was still only Spanish ships that came and went in Mexican ports until 1821. In three years numerous North American and European ships were anchoring there, but none from Spain.

When all Mexican ports were finally opened to trade with the world in 1821, there was still an interval of some confusion in Veracruz—an anticlimax to Independence. Between 1821 and 1825 the fort of San Juan de Ulúa just offshore in front of Veracruz was held by the Spaniards, even though they had already lost the mainland. It took a year or two for the Spanish merchants to wind up their affairs in Veracruz and for others to set up operations in their place. Moreover, government of the town itself was contested during that time by the forces of independent Mexico's first head of state, the emperor Iturbide, and the Republicans under Generals Santa Anna and Victoria, all of which did not do a great deal for trade. In fact, during this interval, as during other times in later years when French or American ships blockaded the harbour of Veracruz, most trading shifted to Alvarado, a small coastal town some seventy-five kilometres to the southeast, at the mouth of the Papaloapán River.

Foreign penetration through Veracruz was attempted nevertheless. An English ship, the *Rawlings*, cautiously entered the harbour of Veracruz on 3 March 1823. It had been chartered by a German stock company, the Rheinisch-Westindische Kompanie. The four German merchants who accompanied its shipment of linen and various luxury articles are thus to be credited with ushering in the new trading order (Pferdekamp 1958:33). The crew and passengers could see the Spanish flag on the fort but were quite uncertain into whose hands the port or indeed the country had fallen. Only through communication between ships' captains was it established that the Republicans and General Victoria controlled the town.

H.G. Ward, who was eventually to become the first English chargé d'affaires in Mexico, came into port later the same year. His party was feted in the house of General Victoria, the man soon to become president and already most desirous of contacts with Britain. Ward had just been escorted through an empty town to the boat that would return him to his ship for lodging overnight when his Mexican guard of honour decided to fire a salute. Unfortunately, they had forgotten that their guns were shotted and pointed in the direction of the Spanish-occupied fort; a cannonade ensued back and forth over the Englishman's head (Ward 1828:II, 176).

The confusion and dislocation consequent on the final Spanish tenacity in the fort of San Juan de Ulúa actually lowered trade out of the Gulf Coast ports quite drastically after Independence. Before the end of 1824, however, it was back up to pre-Independence levels; by 1826 the volume had increased well beyond the old levels, and by mid-century had almost doubled. The *apertura* was providing tangible benefits.

Ward was to reflect, eventually, that "it was, indeed a new epoch in the history of America that commenced with our arrival. It was the first step towards that growing intercourse with Europe, the importance of which to them, and to us, will be every day more generally felt" (Ward 1828:II,190).

In fact, of course, Mexico's "new epoch" came to be characterized by nothing so much as stagnation and indeed decline, certainly in the tropical lowlands. It may be sampled in Bushnell and Macaulay's recent book,

The Emergence of Latin America in the Nineteenth Century (1988: 55–82). It has often been called an *época negra*. Both Veracruz and Jalapa had been reduced in population and trade from what they were before Independence and experienced no great upsurge in the succeeding decades. Ruined buildings and ruined streets were noticed repeatedly by authors in the 1820s and 1830s, and later too. Insurgents and foreign invaders at intervals devastated and redevastated villages along the Veracruz-Jalapa road. The reports of French consuls in the 1850s and 1860s are graphic in their descriptions of disease, thievery, high prices and the difficulties of doing business (Diaz 1974: e.g., 29, 53, 64, 65). The road itself was one long ruin. It had been constructed on the eve of revolution with capital put together by the Consulado of Veracruz. It was an impressive piece of engineering with a surface of cement and stone, but it could not be kept up. As for the political situation, historians have had to search their vocabularies for adequate terms, the Mexican historians most of all. Trens, to whom we owe an absorbing history of the port of Veracruz, called it a sad time, bloody and anarchic, a time of cynicism and conspiracy, a time when words like "liberty," "honour," "democracy," and "justice" meant nothing (1955:75).

Such a period does not leave orderly or very comprehensive records and has in fact received little systematic attention. However, there is the potential for more, and it lies to a large extent in the exploitation of the material basic to this analysis (Florescano 1977:435). There has been some close inquiry into the foreign trade of the time, resulting in an elaboration on such summaries of the epoch as that of Trens (1950). Foreign trade was inhibited by "political instability, constant budgetary scarcities, administrative disorder, the struggle to centralize power, the continuous uprisings against central and state authorities, foreign interventions, contraband, the venality of officials, the lack of security personnel in ports and customs houses, the constant changes in tariffs, brigandage, etc." (Herrera Canales 1977:6, my translation). At the same time, therefore, that Mexico made itself freely accessible to foreigners, it also could not avoid offering them a humiliating spectacle.

Since the majority of foreign visitors coming to Mexico during the decades that followed Independence entered the country at Veracruz and proceeded inland along the same road, a certain parallelism was imposed on the resulting travel accounts. This makes it possible to aggregate the commentary relating to successive stages of the many individual journeys rather neatly. The parallelism was disturbed by the initiation of the construction of a railway that followed an alternate route, via Orizaba, to Puebla and Mexico City. By that time the fascination with Mexico had in any case diminished. Information about the country was widespread, investment prospects there had long since been recognized as something less than fabulous, and travel into the country had become commonplace.

The end of this epoch was even less clear-cut than its beginning. With the intimations of new routings for trade and travel, an old rivalry again came into play between Jalapa and Orizaba, key towns on the two roads that had competed for the traffic up to the plateau since early in the colonial period. They had always had a roughly similar potential for industrial development: each had water power in its vicinity and access to regions of agricultural productivity which was important especially with respect to cotton. Both also faced similar problems: mainly the high costs of moving goods inland, a small local market, and a small labour pool. Nevertheless, the two were to develop quite differently (Rees 1976:103–13, 171–6).

In 1851 Alfred de Valois made an excursion on horseback out of Veracruz by the Tampico gate with his host, the consul of the United States, who was also of French origin. They soon came to the right of way of the new railway which was headed toward Jalapa via the lands of Santa Ana, who had given the concession, and who hoped to profit by it. It seemed poorly built; it was still only several kilometres long, had cost an immense amount, and was already in disrepair. "A very Mexican affair," Valois remarked. The consul agreed, and added that an eternity would be required to finish it, as well as the resources of all the mines of the Americas (Valois 1861:58).

The town council of Córdoba, however, was already alarmed about the potential competition. In several years the influence of financial interests in Córdoba and Orizaba had effected a redirection of the railway. The German traveller, Baron von Müller, describes how he was taken out along this newly extended and rerouted line in 1856 on a somewhat comical but quite enjoyable hybrid conveyance, a railway carriage drawn by mules. They could go as far as Tejeria, the end of the rail, which was about thirteen kilometres from the plaza of the port (1864:203). Edward Burnett Tylor had come in a similar way that same year (1961:24). By 1866, another traveller reports, the route had been extended to Camarón, some seventy-three kilometres out of Veracruz, and was now used by locomotive-drawn trains (Domenech 1922:30). The line from Veracruz to Mexico City via Córdoba and Orizaba was complete in 1872. A line was built from Veracruz to Jalapa too, by 1875, but it did not reach Mexico City until 1891 and was unable to counter the baleful effect on Jalapa of the other route. Industries were developing at Orizaba, whereas Jalapa remained a regional agricultural centre.

Most travellers now went into the interior via Orizaba, but it was not only the route that had changed. Many of the constraints of the stagecoach or the litter had been removed. Travel was faster and the stops were fewer. Subtle environmental changes once experienced in detail, kilometre by kilometre, were less remarkable. In 1889, a German traveller was pleased to recall that he had known Mexico when it was still only lightly affected by the change in transportation technology. For him the coming of the rails had extinguished the romance of travel (Hesse-Wartegg 1890:iv).

In addition to the common route and means of travel, various strands running through the accounts themselves unify the commentary made by observers over the period of nearly a half a century after independence. The foreign authors made many explicit and implicit references to each other. Some authors writing around mid-century were still referring with respect to what was seen by the earliest post-Independence observers; indeed, several of them were presuming only to enhance what Humboldt had set down at the turn of the century. There is no evidence within our body of literature for a change in paradigm during the years under consideration. Nature and man were seen in roughly comparable terms throughout. Moreover, there seems to have been no need for any one of the observers to say in effect that in the early years after Independence it was thus and so in the lowlands but that since then things had changed substantially. Historians of modern Mexico, either Mexican or foreign, do not usually disaggregate the first half of the nineteenth century, certainly as far as the lowlands are concerned. For a reasonably educated European or North American of the second half of the century wishing to inform himself, say, regarding the context of the installation of an emperor in Mexico in the 1860s, those from our list of accounts that would have been available to him would probably have melded into one picture, as the whole series does when read now. Ortega y Medina's theatre troupe does indeed perform as a group.

We have long ago accepted O'Gorman's dictum that early America was not so much discovered as it was invented—the great new option for Western man (O'Gorman 1961). The observers did not simply see what was there, but what seemed or needed to be there. America was thus given a meaning external to itself. The tropical lowlands of the new republic of Mexico were similarly invested by the early nineteenth-century travellers. Dunes, swamps, forests and free-standing oaks were assigned their significance. Imagery from Holy Scripture, classical antiquity, or the Romantic poets was pressed into service, together with such original imagery as the travellers themselves had the wit to confect.

There is factual information in these travel accounts. There are lists of plants seen along the road or in the gardens, of the occupations represented among the working population of Veracruz, the numbers of people, male and female, to be found in rural and urban entities, and a good deal more. Such information will be used in this analysis; the imagery, however, will be taken as the more important data. The way in which this landscape was seen, made plausible, coloured, moulded and envenomed for a wide and avid readership is the main concern. Facts versus imagery: it is interesting to see how the observers themselves evaluated their material in this regard.

APOLOGETIC PREFACES

It is common, of course, for authors to limit their liabilities, but among this company there seems to be an extraordinary defensiveness. This is particularly true for the diplomats, the commercial travellers, the tourists, and the journalists, somewhat less so for the naturalists, who seemed confident that their scientific interests largely recommended themselves. The military authors extended their own particular kind of apologia throughout their volumes. Most of the authors of the systematic treatises validated them in the typically academic way, with an introductory survey of the sources.

There are somewhat ritualized references to sources scattered throughout the literature under study. Personal observations, several authors maintained, had been reinforced by consultation of literature they considered reliable. Here is an example: "I have taken some pains in collating the lucubrations of most of the writers on Mexico who have preceded me; and where I have been sure their information was correct, their descriptions faithful, and their observations just, I have not hesitated to reproduce their corroborative testimony, interwoven with my own" (Robertson 1853:I,viii).

Such literature might or might not be cited with sufficient detail so that it could actually be traced; if not, then that was only so that the text would not become more cumbersome than it already was. The novelist Sealsfield (1974) simply pointed out, in effect, that although his sources were certainly reliable he had assumed no high purpose and hence did not feel obliged to cite them.

Some authors were disarmingly frank. One admitted he was recording his travel recollections because he had some idle time (Elton 1867:iii). The ex-President U.S. Grant was aging, in difficult financial straits, and confined to bed by a bad fall when an unnamed friend suggested that he write his memoirs. "I consented for the money it gave me" (Grant 1952). It is astonishing to see how many writers initially had no intentions of publishing their material. Evidently this must be said even of Frances Calderón de la Barca, whose letters written to her family were "*really,* not intended originally— however incredible the assertion—for publication . . . I strongly recommended that they should be given to the world" (William H. Prescott in Calderón de la Barca 1843:xxv). A clamouring public sometimes made publication imperative. In England in the 1820s, to hear William Bullock tell it, anyone with recent new information on Mexico was besieged and had to publish in self-defence. For other authors, there was the logic of the sequel: "During the last year I published a few voyages to various parts of the world, which were so well received that I am induced to narrate a second series" (Coggeshall 1858:7). Some authors maintained they wanted only to amuse; others wished both to amuse and to instruct. One writer wanted to drop some hints about antiquities in Mexico so that Americans more qualified than he might be stimulated to do some research there—otherwise the Europeans would get it all (Norman 1845:vi). It might even be that if one awakened some interest in the condition of Mexico, this might "tend towards its amelioration" (Mason 1852:viii).

There was sometimes a final unavoidable rush and the material had to be surrendered to the printer as it was. There was no time or opportunity to do a great deal of additional research for statistics, to check spellings or to include the reflective, synthesizing passages that would have put the discussion on to a higher plane. Better to short-circuit eventual criticism by protesting that one had no pretensions.

There were occasional reflections about the malleability of perceptions. Madame Calderón de la Barca, for example, noted the difference between her first and her later impressions. In the interval there had been conditioning, enough contacts with Mexicans to allow a more sympathetic interpretation, perhaps, and some reflections on what might be politic or "correct." More often, returning travellers passed over what they had already seen and described. Most of the raw material of this book is in fact made up of first impressions. Several authors remarked that they wished specifically to preserve the immediacy of their observations, to have them printed much as they were written. They realized that to tamper with the journal entries was to dull them. They also seem to have realized that to attempt to set their accounts into a context and generalize from them would commit them to a larger, different, and much more demanding exercise for which they would probably not be prepared.

Mostly, the accounts were affirmed to have been set down without predisposition. The authors were reporting only what they had seen or knew from reliable sources to be true. There seems a lingering consciousness here of the common scepticism that had prevailed for centuries regarding travel books: they were often considered "repositories of wonderful lies" (Fussell 1980:165). Swift had satirized them in his *Gulliver's Travels* (1726). The nineteenth-century travellers leave the modern reader nostalgic for the days when objectivity was believed to be a realistic goal. There were many ways of saying this. Becher (1834) quoted Shakespeare on his title page: "and by your gracious patience/I will a round unvarnished tale deliver" (*Othello* i.iii).

Such protestations undoubtedly reflect the prevailing enthusiasm for science, but they also indicate a lack of self-consciousness over bias. We do seem to be faced, therefore, with unguarded material, especially in the observations made in the lowlands, which were simply along the way to other places where the main interests of the travellers lay. All of this is highly promising for an analysis of predispositions.

CHAPTER TWO

Nearing the Coast

Travellers approaching the coast from the east knew that they could expect a striking first view. "Every telescope was in requisition" (Bullock 1825:1,6). Sometimes the clouds remained opaque, but in better times they would part, perhaps withdraw completely, and reveal Mount Orizaba, a snow-capped peak southwest of Veracruz. It was a dormant volcanic cone 5,295 metres high, visible many kilometres out to sea—the first of a sequence of signals that the landscape would give them as they travelled into Mexico. In one account after another, one senses a focusing of attention. The sight of the peak was a thrill and a relief. As the landscape at its base cleared, there was the band of green hills that promised tropical luxuriance; soon travellers would see the intervening dunes and the port as well. It was time now to look for the first confirmations of what one had read and heard regarding this country and its people, to face the hazards of a tropical coast, to make the first Near Eastern allusions and, for many, to begin the personal romanticization of this country. Later, on departure, El Pico would be a sign again. Its disappearance below the horizon would allow the travellers to turn their backs on Mexico.

The tone of the journal entries varies. Passengers of strong religious inclination reflected on the magnificence of the work of the Creator. Some merely registered a banal remark about the mountain's beauty; for others its appearance precipitated an emotional passage:

It rose beyond mountains so far off that all trace of chasm or ledge or belting forest was folded in a veil of blue air, yet its grand, immaculate cone, of perfect outline, was so white, so dazzling, so pure in its frozen clearness, like that of an Arctic morn, that the eye lost its sense of the airy gulf between, and it seemed that I might stretch out my hand and touch it. No peaks among mountains can be more sublime than Orizaba. Rising from the level of the sea and the perpetual summer of the tropics, with an unbroken line to the height of eighteen thousand feet, it stands singly above the other ranges with its spotless crown of snow, as some giant white-haired northern king might stand among a host of the weak, effeminate sybarites of the south. (Taylor 1850, 11:324-5)

Some travellers were prepared to entrust more to their journals. Waddy Thompson, an American diplomat arriving in 1842, penned an intimation of his country's expansionism at the sight of *El Pico*:

I can conceive nothing which conveys more of the sublime and beautiful than this lofty mountain ... seen from on board a ship of war; a union of all the grandeur and sublimity of a lofty mountain with the vastness and power of the ocean, and the symmetry and beauty of one of the noblest structures of man. (1846:2)

Such introductory elation was soon counterbalanced. Sealsfield, an American novelist, noted in his journal that for him the approach had evoked contradictory sensations. He had seen the flaming pyramid, but then had soon become aware of the white forms of Veracruz itself as well, and these seemed whited sepulchres. Similar thoughts occurred to many others, here during the approach and later on. Apprehensions, outright fear, disapproval, even revulsion, offset attraction. The evaluation of the country, certainly its coastal lowlands, would reduce itself repeatedly to countervailing propositions.

Between the first sighting of Mount Orizaba and the going ashore there was usually a considerable interval. The wind could die down, the pilots or customs officials might hesitate to come out, nightfall might intervene, or, more seriously, a storm could blow in very quickly and send the sailing vessels, at least, scurrying away to ride it out safely, clear of the shallows that ring the harbour of Veracruz (Figure 6).

RECOGNITION

There was usually ample time to study the shoreline and the surroundings of the anchorage, and time to write in journals. Together these yield an almost surrealistically detailed view of whatever the travellers could fasten their attention on while waiting. They seemed eager to see their first Mexicans and to verify what they knew of them. Several looked closely into the faces of the pilots and watched their gestures, ready to render, even before the anchors were lowered, a few first characterizations (e.g., Gilliam 1847:20). While his ship lay at anchor,

6 Bird's eye view of Veracruz, c. 1850, by Casimiro Castro as in Pourade (1970:31)

Becher scanned the passengers in the service boat that plied between the quay and the fort of San Juan de Ulúa twice each day (Becher 1834:33). He saw a range of skin tones, which bothered him, and many very scantily clad people. One could see, he ventured, how European clothes were not quite suitable in these temperatures, but this nakedness would take some getting used to.

San Juan de Ulúa

Every visitor arriving at Veracruz had to notice the old fort that guarded Mexico's eastern approaches (Figure 6). It stimulated historical reflections, and although most travellers brought with them a critical view of the legacy of Spain in the New World, this fortification elicited respect. Artists had painted and sketched it, attempting to give it some presence and a few bold lines, but it is just not a very imposing structure, and neither are most of the other forts of the New World. They worthily commemorate the legacy of Spain, but they have little beauty, as a Spanish historian has admitted, little of the medieval castle's power of romantic evocation (Diego Angulo Iniguez in prologue to Calderón Quijano 1953:xi).

The San Juan de Ulúa seen by the travellers had been given its form by a capable group of engineers working during the 1770s and 1780s on the improvement of the defences of the Gulf Coast in preparation for an expected attack by the English. Prominent among these was Miguel del Corral, who had lived and worked in the Gulf lowlands for seventeen years and who left a remarkable description of southern Veracruz in the 1770s (Siemens and Brinckmann 1976). He contributed to the design of the whole system of fortifications that was to bar the eastern entrance of New Spain. It included a series of coastal gun emplacements intended to deny the English a landing on the beaches to the south of Veracruz, and to protect the mouths of the Papaloapan and Coatzacoalcos rivers. The town wall was fortified and a fort erected, as well, near the summit of the road, at Perote. This was to allow a last stand in case of invasion and to provide the defenders with a base at a healthy elevation for forays into the lowlands. The key to it all was San Juan de Ulúa.

An English traveller remarked that "The castle of St. Juan de Ulúa belongs to that class of fortresses whose real strength is much more remarkable, than their outward appearance is striking or picturesque. It lies low on the water, in the midst of the harbourage, having for its base nothing more elevated than a mere sandbank, of which the shallow flats form its defense to the sea-ward" (Latrobe 1836:300–1).

The fort was built of fossil coral from nearby reefs, called *madrépora*, or *piedra múcara*. Its walls absorbed the shock of an incoming cannon ball and did not easily shatter. Many travellers remembered that it had been heroically defended by the Spaniards, particularly from 1821 to 1825, when it represented their last hold on New Spain. The somewhat-less-than-heroic later surrender of the fort by its Mexican defenders, after a short bombardment by the French punitive expedition in 1838, was also noted. Madame Calderón de la Barca expressed what many conceded, that the fort remained "a lasting memorial of the great works which, almost immediately after their arrival on these shores, were undertaken by the Spanish conquerors" (Fisher ed. 1966:610).

By 1847, San Juan de Ulúa had been extensively repaired and its armament doubled. General Scott, the commander of the invading American army, maintained that "When we approached, in 1847, the castle had the capacity to sink the entire American navy." The guns remained inexplicably silent during the invasion but soon thereafter opened fire, and Scott had to take pains to arrange his troops in such a way that the town was between them and the "terrible fire of the castle" (Scott 1864:422–6). It might have been used to much greater advantage still, and this was a key point. Passing through the port a year after the war, Taylor, the talented correspondent of the New York *Tribune*, saw it as "a place of immense strength, [that] in the hands of men who knew how to defend it, need no more be taken than Gibraltar" (Taylor 1850:330). Taylor was echoing an earlier and perhaps more sensitive judgment by Eugene Maissin, chief of staff of Admiral Baudin, commander of the French blockade of Veracruz: "The location of the fort is strategically sound; its fortifications are well planned; its casements are bomb-proof. More skillfully defended, the fort would not have fallen so easily into our hands. The Mexicans who manned it were courageous and fought for a long stretch of time; but they lacked skill and also the steadfastness that will not surrender until all means of defense are exhausted. They gave up before they had to" (Maissin 1839/1961:74).

No matter how much one might sympathize with the struggle in the southern republics to throw off the yoke of Spain, its practical results represented by this misused fort—and the ruined road they would soon see—had been unfortunate.

Isla de Sacrificios

The best anchorage was in the lee of the fort, but the Isla de Sacrificios offered some limited shelter as well. It had emerged, like the land on which the fort was built, from one of the ring of shallows around the port. Tayloe, the young Virginian on Poinsett's staff, saw his first Mexicans there. At the time, the island was in use as a penal colony for culprits from all over Mexico. "The sight of these miserable creatures,

and the ill-disciplined and ill-looking soldiers who guarded them, gave
me a most unfavorable impression of the Mexican nation. All the troops
are composed of mestizoes—a breed of Spanish and Indian mixture"
(Tayloe in Gardiner 1959:18).

Sacrificios, a Prehispanic burial ground and place of ritual, is briefly
illuminated in the travel literature by the attention of some of the personnel of the French punitive expedition of 1838–39, whose ships were
anchored in its lee (Blanchard 1839:63–76). There was little more than a
low beach and some scraggly vegetation, but it was all the men had to
absorb their attention, since most were unable to go ashore during the
months of negotiations. It was a place to bury those who died of yellow
fever. There was neither produce nor water to be had on the island; on
occasion a ship had to be sent to Havana for fresh supplies. Cattle could
be pastured as live provisions. One could take a ten-minute walk around
the beach; the sailors occasionally engaged in "bull-runs." All in all, it
ill rewarded the men, virtually captives on their own ship, for the long
crossing. They, like everyone else who made the journey, expected tropical luxuriance. Instead, the island, and indeed the mainland, presented
"inhospitable shores, whose monotonous and silent outline depressed
the eye" (Maissin 1839/1961:123).

Veracruz

The melancholy aspect of the town and its desert-like surroundings,
seen from offshore or from inland, was frequently described. At first
sight the town itself did not merit this sombre opinion: "Its red and
white domes, cupolas, terraces, convents, hospitals, churches, etc., with
the regularity of its walls and batteries, had a splendid appearance from
the water; but alas! it is but a painted Golgotha, the headquarters of
Death, for I believe it to be one of the most unhealthy spots on earth!"
(Bullock 1825:I,13–14).

The town had been rendered macabre by yellow fever. The dunes
around, cheerless enough for their lack of greenery, were often impli-
cated in the generation of this menace. A stay in the port was always
overshadowed and the departure from it not infrequently hastened,
especially during the rainy season, by the danger of infection. Many
journal entries made during the visit to Veracruz, which more than one
traveller called a detention, included reflections on the origin and
nature of this disease.

A few were able to take a more detached view. The well-read English
gentleman traveller, William Parish Robertson, who, having paid for the
publication of his memoirs, could indulge himself on many topics, took
deliberate issue with the reputation of Veracruz and its surroundings:

As we approached Vera Cruz, we had a fine view of the coast. . . .
Expecting to see interminable sand-banks on the coast as far as the
eye could reach, we were agreeably surprised to find that they only
extended something less than two leagues on either side, beyond
which the country was green. . . . The town also looked a great deal
better than we had expected to find it. Indeed it had a handsome and
imposing appearance; but with San Juan de Ulloa, of which, among
Spaniards, I had heard so much, I was quite disappointed. It presented a black, dilapidated, ruinous look, without any of the frowning
grandeur with which I had invested in my own mind, this celebrated
fortification. (Robertson 1853:1,231–2)

He admitted the danger of yellow fever—for others. He was confident
that he had come during the healthy season and, moreover, he was a
rational man, not given to excesses, and hence quite safe. Others were
unable to muster his detachment: "The first object that attracted our
attention was the lowering of two dead bodies in coffins from the sea-
wall of the castle, into a boat, and the next a pile of empty coffins on the
mole as we landed from the ship to report our arrival. Everything else
was dead in the blazing, glistening sunlight; not a living object, not one
moving thing could I see . . . to tell me that the yellow fever, the dread
black vomit, was raging in this fiery oven of a plague-stricken city"
(Kenly 1873:286).

Stared at long enough from over a ship's rail, the coastline could
become an abstraction. One traveller found that the town had no depth,
only a long, low facade (Buschmann in Thümmel 1848:101). A cardboard cutout emerges from the other descriptions too: a row of buildings, with the famous scavengers of Veracruz, the vultures, hopping
about on the roofs, which in turn appeared as a row of "monumentos
funerarios" (de Fossey 1964:249). All was "grey stone, only relieved by
the coloured Spanish tiles on the church roofs, and a flag or two in the
harbour. Not a scrap of vegetation to be seen, and the rays of tropical
sun pouring down upon us" (Tylor 1861:19). Figures swarmed over the
quay; others animated the one transverse street along which one could
see into town. A strip of sand extended to both sides, punctuated by
wrecks. One could just make out the traffic along the beach to the
northwest, the first stage of the road to Jalapa. Behind all this lay
undulating green, compensating the viewer somewhat for the barren
foreground; then hazy dorsal spines of mountain ranges, and the peak
of Orizaba. At night it all reduced itself to a line of twenty lanterns.
Occasionally during the spring burning season the hills behind the
town would light up as well.

Anyone on deck early in the morning would hear military music from the fort and then a cannon shot, assuring all that the defenders were on the alert. During the day, if the wind was offshore, there would be a repeated ringing of bells. On days when the wind was in the other direction it was quiet.

The view from the deck stimulated allusions to the Mediterranean and the Near East in at least a dozen accounts. Müller, who had travelled in the Orient, was moved to remark that he had seen all this before, the domes and flat roofs, the Christian towers that suggested minarets (Müller 1864:193). The sand around the town was an Arabian desert (Calderón de la Barca, Fisher ed. 1966:53). Behind the sand and the town, a Mexican traveller noted almost as an apology, lay the real Mexico—a beautiful face behind a veil, the country was as a mansion in Alexandria: behind its ruinous facade one glimpsed sumptuous rooms (Payno in Tavera Alfaro 1964:110).

The French traveller Ampère made a more fundamental Egyptian reference. Veracruz was to yellow fever as Egypt to pestilence (Ampère [1856] in Glantz 1964:413). Jehovah had brought down ten plagues on Egypt as a punishment for a restrictive emigration policy. The water of the Nile had turned to blood; frogs, lice, and flies infested the land; cattle became diseased and then the Egyptains themselves. A hail storm raged over the farmland. Locusts and darkness descended; the final horror was the death of the Egyptian firstborn. All this was sent down on Egypt as a judgment; the Lord even hardened Pharaoh's heart to increase the duration and effect of the punishment, and thus Egypt was left with a pestilential image. To transfer the image to Veracruz was to apply one of the strongest stigmas in the Judeo-Christian heritage.

APPREHENSIONS

There was more than enough time before disembarkation to reflect on why one had come and the dangers that lay ahead, to rehearse and exchange in worried tones information on the weather, the road, and the town, to contemplate the possibility of death by yellow fever before having had the chance to get inland. This was a sinister, mysterious evil that one could only hope to avoid by some good fortune. It was well known also that there would be terrible inns and food, a frightful shaking up in the stagecoach and a variety of other indignities on this journey into Mexico. Returned travellers had been spreading such information with relish for many years, speaking of the country in deprecating terms, "hablando pestes," as one Mexican author put it (Payno in Tavera Alfaro 1964:106). In preparation for a journey to Mexico, as well as after the return, and indeed along the Mexican roads themselves, one of the staples of discussion was always some tale of robbery. One could come to terms with this threat; one could arm oneself against thieves.

Carl Heller recorded his apprehensions in these respects particularly well during his wait at anchor. On the one hand was his thrill at the prospect of an entry into the tropics, his hope for good new data, a hope familiar to an academic going into the field. On the other was his unease over unfamiliar customs and language, over arbitrary official-dom and outright lawlessness. He would be dependent on luck and his own wit (Heller 1853:33-4). He was thus poised somewhere between fascination and dread, as were most travellers to this coast. This tension, perhaps more than anything else, stimulated the rich commentary.

El Norte

Between October and April, storms from the north could spring up with short warning. They made it extremely risky to attempt a penetration of the ring of shallows that surrounded the port. Koppe and his fellow passengers were sobered on their approach to Veracruz just after a *norte* by the sight of a shipwreck in one of the shallows; the passengers and goods were just being cleared away. When they arrived at the anchorage there was the wreck of another ship; it had been torn from its moorings in the lee of the fort and smashed against the city wall (Koppe 1955:50-1). One could, in fact, never be sure one was safe from this menace until one was actually on land.

Humboldt had tersely formulated the basic alternatives for anyone contemplating a journey to Mexico, and they had become general knowledge (Humboldt 1809, 1:71). There was a simple choice between the principal hazards, and thus of timing: yellow fever, highly likely during the rainy summer months, but possible all year, or the *norte*, which struck at uneven intervals and with uneven intensities during the winter. These same hazards, plus the tremendous topographic incline on the margins of the Sierra Madre Oriental and the bars at all the river mouths, had been New Spain's first and quite effective eastern line of defence. They imposed additional costs on the ramified post-Indepen-dence contacts with Europe and North America as well, but could not impede them.

Most travellers chose the time of the *norte*: the odds were better. Most journals therefore include entries on the perception and experience of this meteorological event. It did not cast a pall over the port or the journey inland as did yellow fever. Rather, it added to the already con-siderable shipboard anxiety the special agitation of a serious gamble. The best of luck was to arrive between *nortes*:

So, it is in fear and trembling lest a *norte* should spring up, and snatch away his chance of getting ashore, that the traveller passes the hours on shipboard which intervene between sighting the coast and anchoring in the roads. For the *norte* may spring up when the sky looks least threatening, and woe be to those who are caught by its violence. Then the traveller, who has been rejoicing in the prospect of getting ashore in a few hours, must resign himself with the best grace he may to being tossed by merciless waves until the *norte* shall have spent its fury—it may be in two or it may be in fourteen days. (Bullock 1866:2)

The phenomenon is easily summarized in contemporary climatological terms (e.g., Garcia 1970:8). It originates in the build-up of a polar continental air mass, a high pressure system, over the North American Rockies and their eastern slopes. Cold air, with a front along its leading edge, breaks out of such a system in a southerly or southeasterly direction. Over the warm Gulf waters it gains in temperature as well as humidity and becomes unstable. The results include a gusty north wind, clouds, showers, and seas that may run to ten metres. The front may develop a wave, that is, become cyclonic, which accelerates the movement southward.

As the front approaches any given point a cloud bank forms to the north. As it passes, the temperature drops quickly, winds freshen from the northwest or the north and the pressure rises again. If the front has developed a wave, and this passes over a point obliquely from northwest to southeast, a succession of further wind shifts and variations in land temperatures may be expected. As many as thirty such disturbances may be expected in the course of a dry season. Perhaps ten of these will develop gale force winds, which are likely from December to February.

Next to such formulations the early nineteenth-century descriptions seem impressionistic and excessive in their nuance, but much better reading. In the weather observations, as in the discussion of the causes, course and treatment of yellow fever, one sees a groping toward the generalizations that are common knowledge now. One also finds intimations of research problems still unresolved.

The *norte* was described in detail by six of our authors. Ortega y Medina, with a certain amount of sarcasm, noted how nineteenth-century travellers just could not seem to desist from describing a few sea storms: they were perfect subjects on which to express romanticism (Ortega y Medina 1955:17). Humboldt did it with detachment, synthesizing the best information available. Mühlenpfordt had evidently not yet found anything better forty years later, and paraphrased him point for point. Latrobe was genuinely frightened and also revolted by the shipboard conditions under which he had to experience his *norte*. Ferry wove weather observation into what may have been a fictionalized account of the work of wreckers. Blanchard was at first curious and then thrilled by the spectacle. Madame Calderón de la Barca was disgusted with the shaking up she sustained and the delay.

Those coming into port or already at anchor were understandably sensitive to any signs of an impending *norte*. Crew members would point them out or the travellers might recognize them from what they had read. Swarming insects, fish leaping out of the water, sea birds wheeling round and round in the air uttering mournful cries: all these could be signs. Winds and breezes were read this way and that. A common ditty embodied one interpretation: *Sur duro,/Norte seguro*. (A south wind strong,/The norther ere long.) (Calderón de la Barca, Fisher ed. 1966:46). Such a south wind could be moving into a low pressure area over the Gulf; it is an indicator still heeded. Calm, oppressive heat and lightning in the evening were ominous as well. Any clouds that might appear would be carefully scrutinized. Latrobe noted the characteristic wispy high-level clouds that indicate a cyclonic disturbance, one way in which the leading edge of the outbreak can appear. He saw man-of-war birds set out to sea and pelicans fly to the safety of the beach (Latrobe 1836:9–10). A dark cloud on the northern horizon meant that the *norte* was about to strike. Humboldt identified an extraordinarily heavy dew as a forerunner, and a most intriguing further sign, a sudden clearing of the mountain peaks along the horizon just before the strike. There does not yet seem to be an explanation for this. Humboldt carried a barometer with him on his journeys through the Americas; the mercury became restless at a time like this (Humboldt 1809, 1:70).

Madame Calderón de la Barca noted that as the storm approached the face of their first lieutenant "fell as many degrees as the instrument" (Calderón de la Barca, Fisher ed. 1966:46). At such a time most captains already anchored opted for flight, even if it meant separating passengers still on board from baggage already sent ashore, or vice versa. One could exploit the ominous south wind to sail in an easterly direction and get well clear of the shallows. If a ship could tie up to one of the stout rings in the southwest wall of the fort, a privileged location, and throw anchors aft, it could probably ride out the storm. This would still mean a shake-up for all aboard, and no communication with the shore for the duration. A steam boat, such as those that begin to appear in the accounts in the 1930s, might use power and anchors to achieve the same end. Anchors alone were unsafe; the wrecks along the beach were quite convincing.

The arrival of the cold wind proper, the *norte* itself, was sensed as a sudden physical blow: "Such a wind I had till then never witnessed. The sea was apparently levelled under its pressure; and far and near seemed like a carpet of driving snow, from the sleet and foam which was raised and hurried along its surface" (Latrobe 1836:15). This was an avenger, descending sword in hand, as inhabitants of the coast sometimes put it (Ferry 1856:318). It was the biblical flaming sword that guarded the gates of Paradise (Calderón de la Barca, Fisher ed. 1966:50).

The ships that had not made it safely out to sea or that had broken free were in extreme danger among the shallows, between a driving wind and a transverse shore. It was all made worse by wreckers. These would attempt to mislead ships with beacons in order to get them hard aground within range. They would probably rescue the crew and passengers, but would also get what they could from the wreck before it disintegrated, and would certainly collect what was washed ashore—a bounty from the sea. Boca del Río, a village just south of Veracruz, was excellently placed for this sort of thing. As a last resort a ship might try to get over the bar at the mouth of the Jamapa River nearby, with the help of a high sea, and thus reach shelter. It could be diverted as easily to the one side as to the other.

Passengers on ships riding out the *norte* could be pushed to the limits of their endurance, which resulted in some fairly good prose: "What between the crowded state of the cabins, the violence of the storm in which we were involved, the fears and terrors of some, the horrid and blasphemous language uttered by others of the desperadoes about us, the dirt and impurity surrounding us, and the quarrelling and caballing of the crew, our position was truly unenviable" (Latrobe 1836:16). Blanchard describes with what seems amusement the agility required to eat under these circumstances, especially if the cook had succeeded in preparing soup (Blanchard 1839:58). Madame Calderón de la Barca happened to be detained from landfall by a fickle *norte* that first threatened, and then blew on and off for sixteen days (Calderón de la Barca, Fisher ed. 1966:46–51). On the seventh day she described "a night that seemed interminable, we were tossed up and down, knocked against the furniture, and otherwise maltreated." The time was not half gone and she had read almost everything she could find on board. "I am now reduced to a very serious Spanish work on the truth of Christianity." On the fifteenth day she was speechless: "I shall fold up this seascrawl and write no more till we reach Vera Cruz." Fortunately, the wind changed during the night. In the morning an onshore breeze, in the normal pattern of the trade winds of the season, brought them into port.

CHAPTER THREE

The Macabre Port

When it was finally time to disembark, the distant landscape that promised tropical luxuriance, the dunes, the tiresome anchorage and the cardboard cut-out town itself receded from the traveller's view. All attention was now on the immediate: the clamber into lighters and then onto the quay, baggage inspection and the importunities of the *cargadores*. Some spared a glance at the Mexicans who had come to stare at the new arrivals; all were intent on looking up their contacts, finding accommodation and making preparations for the journey inland. And everyone knew that one could easily die of yellow fever while in this town, even in the off season, because in some years it persisted through the drier months. So after weeks of lethargy, hurry had set in. No one would breathe easily again until they had crossed the lowlands and seen the oaks on the slopes just below Jalapa.

Thus Veracruz was viewed in passing; it is not described very deliberately by most of the observers. In addition, each one of them had contacts within only a limited segment of the town's population; most spoke no Spanish. The town emerges more comprehensively, of course, when one pieces together the foreign commentary of the first half of the century. Even so, however, one senses substantial lacunae. There is little about the life and circumstances of the majority of the population. In fact, behind the travellers' descriptions, which themselves include several distinctive perceptions of the town, there is the suggestion of another Veracruz, which obtruded on first-hand Mexican accounts of the time.

Can one meld the commentary made by many observers over a period of about half a century into one town? There were special circumstances that affected what some of the observers saw, such as the various sieges and evacuations. Aside from these events, the main economic activities and conditions in the town, its total population, the racial and ethnic composition, customs and practices, certainly the physical form of the town, all these aspects changed little during the period under consideration. More to the point, the imagery of the town, the lowlands, and indeed the country confected during this time by foreign observers evolved by interstimulation as much as by individual imagination; there is evidence that it was regarded as cumulative by the writers themselves and by their readership, as has been pointed out in Chapter 1.

The travellers had been eager to begin their reading of coastal Mexico in the faces of the pilots and the oarsmen of the lighters; in the port itself they could begin the confirmation of their predispositions in earnest. Most brought prejudices with them regarding the dark skin tones apparent all around, as well as some further ideas about lazy natives and about untrustworthy Mexicans in general. There was ample opportunity, even during a short stay, to find corroboration for some of this right here within the port. Some of the travellers became quite intrigued with the goings-on in the streets and discussed Veracruzan women or the view from the watergate during a *norte* with considerable sensitivity. All in all, then, we are given quite an exotic port, as tropical ports should be, complete with sounds. Much is said about the layout and form of the town; even the outsider who only passed through its streets could obtain some strong impressions in these respects. However, Veracruz was usually seen against a background from another urban tradition. And most foreign visitors, indeed those from the uplands too, sensed the town's frightening, paradoxical aura. In one traveller's mind Veracruz glowed unnaturally (Ruxton 1855:24). No one could actually see why it should be unhealthy, in fact there were reasons why the opposite should be the case, but neither could most visitors ignore that it seemed deadly.

COMING ASHORE

Those travellers that mention their first contact with official Mexico, with health inspectors, customs officials or perhaps a special envoy coming out in a boat, found it all bothersome or arbitrary, or perhaps amusing, and could not avoid a tone of condescension in their descriptions. This same tone may be detected repeatedly throughout the rest of the journey. Koppe sketches a funny scene: the health inspectors who had come aboard his ship are gravely examining the armpits of the sailors for signs of pestilence—here in the paradise of yellow fever! (Koppe 1955:51–2). They were probably looking for signs of cholera, which was being reported as rampant in Europe not long after the time Koppe arrived in Mexico (Becher 1834:30–1). The customs official dealt

rapidly and courteously with Koppe's baggage, no doubt because someone from the Veracruzan office of his company had sent an interceding
word—surely the best way, he felt, to deal with Mexican officialdom.

If it was known that there were dignitaries aboard approaching ships,
guns were fired on their arrival. The first American ambassador, Poinsett, got an artillery salute (Tayloe in Gardiner 1959:18). The first Spanish ambassador, Calderón de la Barca, got a similar cannonade, to hear
his wife tell it (Calderón de la Barca, Fisher ed. 1966:25). In 1846 the
American traveller George Ruxton happened to be in Veracruz when
Santa Anna returned from exile to take charge of the war against the
United States. There were no "vivas" nor any applause, to say nothing of
salvos, since Santa Anna was widely suspected of having already made
some secret agreement with the Americans (Ruxton 1855:29–30).

For Veracruzanos, disembarkations were an entertainment. It is not
difficult to imagine some of the thoughts that went through their minds
and the comments they bandied about between them. There was always
great interest in what the new arrivals wore, a healthy quantum of derision, no doubt, and a good deal of more profound anti-foreign sentiment
too, growing with each armed intervention. Some of the more thoughtful
will have wondered how they were being seen by the foreigners.

Sartorius arranged the typical crowd on the quay into the first of his
many tableaux (1858/1961:1–2). He had a fine way of verbally sketching
the inhabitants of town and country, in their venue and in mid-gesture.
He noticed the wide range of skin tones on the quay and the clothes
being worn: "On the one side Paris fashions, on the other the lightest
possible clothing" (1858/1961:2). To appear cultured in Mexico meant
to wear European clothes. And not just any style would do; one looked
to what was being worn in Paris. This was part of "el afrancesamiento,"
which had been going on in Mexico and elsewhere in the New World
for some time (Miranda 1962:17). It is difficult to imagine now how such
clothes were supportable in the heat and humidity of the tropical lowlands.

The incoming travellers, properly buttoned and stayed, indignantly
censured light clothing, on the waterfront, in the towns, and later along
the road. "Half-naked" was often used, and it connoted barbarism.
Often, however, precisely this barbarism made Mexican women of the
lower classes very attractive. Sartorius could hardly disguise the fact that
he found the loose clothing stimulating (1858/1961:2).

Frances Calderón de la Barca described the crowd that welcomed
their party on disembarkation:

Some had no pantaloons; and others, to make up for their neighbours' deficiencies, had two pair—the upper slit up the side of the leg,

Mexican fashion. All had large Mexican hats, with silver or bead rolls, and every tinge of dark complexion, from the pure Indian upwards. Some dresses were entirely composed of rags, clinging together by the attraction of cohesion; others were nearly whole and had only a few holes to let in the air. All were crowding, jostling, and nearly throwing each other into the water, and gazing with faces of intense curiosity. (Calderón de la Barca, Fisher ed. 1966:54)

AN ALIEN TOWN

There was not much to cheer the foreign visitors as they passed through the watergate and sought out their lodgings, not much, that is, for most of the visitors. But those who would revel in the luxuriance of the vegetation along the road, who carried the instincts of naturalists, allowed curiosity to overcome even the fear of menace that lurked in the town and· looked about them with interest from the moment they stepped ashore.

A sailor who had been to the port repeatedly put it bluntly: "The town has few redeeming qualities, and generally strangers only visit it for the sake of gain, or to pass through it to go to the city of Mexico, and are always glad to leave as soon as possible" (Coggeshall 1858:56).

One of the newly arrived focused on the streets before him the contradiction of the attractive, deadly town: "It is compactly and very well built, and so extremely neat and clean that from an examination of the interior of Vera Cruz, it would be difficult to account for the causes of the pestilential diseases for which it is unfortunately celebrated" (Poinsett 1824:15).

Some twenty years later a fellow countryman pointed out essentially the same thing, adding a magnanimous comparison and some sloppy terminology on disease: "I look at the open sea in front, the exceedingly clean streets, and the desolate coast of sand and stunted shrubbery, which extends north and south as far as the eye can reach. I am at a loss to know why it is so cursed with disease. St. Augustine, St. Mary's and a hundred places among our southern sea-coast, have infinitely more the appearance of nests of malaria" (Mayer 1844:4).

The clean streets of Veracruz were noticed again and again. Urban filth had long been a major preoccupation of Veracruzan authorities. The Conde de Revillagigedo, viceroy from 1780 to 1794, was shocked at the way the people of Veracruz simply threw their wastes into the street (Archer 1971:429). A report laid before the town council in 1797 pointed out that if cleanliness was already regarded as especially important to public health in cities of temperate regions, how much more important in a hot and humid place where the exhalations from sewage and garbage could be so mortal (Trens 1955:46). By the time our travellers were arriving, various measures were actually being taken, no doubt with the odd lapse. Gangs of prisoners, for example, were employed to sweep the streets, pick· up garbage, clear the drains, and remove the sand that drifted in over the walls. In other cities around the Gulf that were menaced by yellow fever, similar efforts were made to prevent the disease.

The travellers were entering a town of the typical colonial American stamp, and opinions varied. Some found it imposing, others tasteless, many dismissed it as just simply bleak and forlorn. In several accounts there is an explicit Anglo-Saxon aversion to the stereotypical grid plan and associated features. This was certainly not the sort of town, irregular in plan, growing up through many individual decisions, such as in response to developments in transportation and trade, that they were then accustomed to. Bullock expressed it best: "He who has seen one town in New Spain has seen them all. There is but one thing to be said in their favour, which is this, that they have the air of having been built to live in and not to be looked at. They are hopelessly regular. The streets all cut each other at right angles, and it is the boast of Spaniards and Mexicans that you can see the country at the end of every street" (Bullock 1866:11).

Casimiro Castro drew Veracruz as it might have been seen from the air about 1850 (Figure 6). It is said to have been done from a balloon, but was probably drawn after a photograph taken in that way (Pourade 1970:72). The French photographer Nadar had been doing this sort of thing over French cities for some years, and it is highly likely that it was tried as another French novelty over Veracruz.

The town seems stoutly defended by a moat, two walls and bulwarks. In effect, however, the main wall was not very substantial and served mainly for the collection of internal duties (Tayloe in Gardiner 1959:20). When sand drifted over it, as often happened during a *norte*, contraband flourished in the town. Otherwise there were three landgates and one watergate. Behind the watergate and in the northern corner of the town were open spaces where goods were transferred and from which mule and cart trains left for the interior. There was, of course, a main plaza, where a market was held each week. To continue the inventory, there were seven churches, four monasteries, two hospitals, five primary schools (for further training people had to go elsewhere), one theatre, two printing presses, and two newspapers: *El Diario*, which is still there, and *La Méxicana* (Thümmel 1848:92-3). The town had gas street lighting by the time Castro made his drawing. The grid of streets, the typically located public buildings, the almost universal courtyard house plan, and the uniform two-storey height are all clearly evident.

Many travellers did not have the opportunity to enter private homes, or public buildings for that matter. They stayed in hotels and left as soon as they could. Of those who did enter homes, most had been invited by foreign residents. A small minority of the authors was able to describe the inside of characteristically built and furnished Spanish-American homes of substance in Veracruz. They were intrigued but found the whole uncongenial. Such architecture is now widely familiar, due to tourism throughout Latin America, and the commercialization of a repertoire of "colonial" architectural imagery, virtually interchangeable between cities. Its elements included an austere facade with barred windows and a large main door into a patio surrounded by arcades. In Veracruz this was where one found a bit of greenery. The ground floor tended to have service functions, the first floor up contained the living quarters. Madame Calderón de la Barca was invited into such a house and found it "immensely large and airy, built in a square, Havana fashion, as they all are, but with that unfurnished, melancholy gaping, which as yet this style of house has to me—(and which makes one feel *triste* and lonely—) though admirably adapted to the climate" (Calderón de la Barca, Fisher ed. 1966:55).

To step from a hot street into the courtyard of a colonial house and then into the *sala* has always been a refreshment. The thick walls provide insulation, there is shade under the arcades. High ceilings and cross-ventilation cool the rooms. All these were noted as useful adaptations, but they did not compensate the visitor with an affinity for a cluttered Anglo-Saxon parlour. Later, in Puebla, the good Madame was invited into another prominent citizen's house and expressed herself more strongly: "It certainly does require some time to become [used] to the style of building adopted in the Spanish colonies. . . . There is something at first sight exceedingly desolate-looking in these great wooden doors, like those of immense barns, the large iron-barred windows, the ill-paved courtyards, even the flat roofs. . . . The whole gives the idea of a total absence of comfort" (Calderón de la Barca, Fisher ed. 1966:82).

Sartorius went into great detail regarding the form and the goings-on in the typical house of reasonably wealthy townsmen. He ended on a similar note: "[The creole] is a friend to luxury, he has showy equipages, beautiful saddle-horses, numerous servants, but no comfort in his house" (Sartorius 1858/1961:56).

The Spanish-American house of substance was found to have two excellent vantage points. The streetside windows of the second storeys opened onto balconies. Koppe found himself sipping chocolate on the balcony of his room on the morning after his arrival. Below him people were going to church and here was the opportunity for his first good go at costumes, something most travellers had to try sooner or later, of course: loose cotton suits on the men, hats, black silk, mantillas, some legs in stockings and some nude, shawls, blouses, delightfully small feet, but not really a beautiful woman anywhere. Passing by down in the street during the daytime, one tried to catch a glimpse of the ladies up above doing their hair, smoking, or just lazing behind the iron filigree screens or under the awnings (Valois 1861:40). The balconies and the hidden women, at least one author felt, added to the oriental aspect already given to the town by its domes and slender steeples (Ferry 1856:315).

Visitors found the flat roof, or *azotea*, an even more engaging feature of the colonial house, especially if it also had a raised, enclosed viewing platform, or *mirador*. One had to be careful, though, because there were watchdogs up there (Koppe 1955:57), as there tend to be on the better houses to this day. At the end of his first day in Veracruz, Müller found himself on a rooftop, enjoying an after-dinner coffee. The subtle colours of the sky, the ships at anchor, the palms, all appealed to his romantic sensibilities, and he too reflected on the irony of this town's frightful reputation among foreigners (1864:1, 195-6).

Such reflections come easily out on the *azotea* and the picturesque is very apparent from a balcony. These travellers were finding within hours of their coming ashore what fine vantage points the traditional Spanish-American house of the well-to-do provided. It allowed a detached view; the sky and the goings-on in the street were an exhibit, especially laid on.

The first night in Veracruz produced a grimier set of impressions for the less well connected. It was the beginning, in this literature, of commentary on the disgusting Mexican hotel. Descriptions of it, suitably embellished on repetition, will have done as much, perhaps, as the tales of highway robbery to form the image of Mexico abroad and to give the prospect of travel there a special piquancy.

Gilliam spent his first night in the Casa de la Diligencia on the main square, the hotel from which the stagecoaches departed for the interior:

It was our determination to spend our first night in Vera Cruz, in quietness and repose, so that we might on the following morning feel refreshed from our sea-voyage, and thereby enjoy our ramble and the view of the city more; but how sadly were we disappointed; for the bells of the public buildings, that were constantly ringing in honor of a saint, as I was informed, perpetuated their clack-a-clack, and we had not more than fairly retired to bed by the hour of nine, than the loud report of a big gun from the castle was heard; then followed the ringing of all the bells of the cathedral and churches, which pro-

duced the most deafening peal that had ever stunned my senses—
this was followed by the beating of drums and the blowing of fifes and
trumpets, and for the balance of the night, as if intended for our
distress, *besides the diligent biting of fleas and mosquitoes* we were kept
awake by the crying of the watch-word of the sentinels, who were
posted at every corner of the street, two of whom had their position
near to us; and the first night, notwithstanding our anticipated luxury
of sleeping in a bed on shore, I was kept listening to the stentorian
voices of the sentinels crying every half hour—Ave Maria purissima,
los dos y media serenis, etc. (Gilliam 1847:21)

The torment of insects was found to be part of most tropical nights.
Even those under mosquito nets would find that there was always one
little horror that somehow slipped under the gauze and circled the face,
humming. Koppe remembered wondering where could be the bellows
in the tiny body to produce such a strident, room-filling sound (Koppe,
1835).

Many spoke of the filth in the hotels, as though our whole company of
authors had assembled in some mythical salon and were uproariously
trying to outdo each other in this respect. One of the best stories must
be Poinsett's account of how a *mozo* who prepared a room for his party
had first to use a shovel and then a broom (Poinsett 1824:25). De Fossey,
the contemplative traveller searching for an opportunity to re-establish
himself after the collapse of the French colony near Coatzacoalcos,
found his tranquillity disturbed by the garbage and vultures below his
window. But he could erase all that by lying down; from his bed the view
lifted to an evening seascape and in the morning, a sunrise (de Fossey
1964: 251).

Mayer had a sensible observation to make about the state of lodgings
in Mexico. Heterogeneous travellers were a new phenomenon in early
nineteenth-century Mexico. Before that, travelling people of the better
sort lodged with acquaintances or in homes to which they had been
referred. Muleteers, who made up the majority of the travellers, needed
only a minimum of shelter. One could not therefore expect the hotels of
Europe along this road (Mayer 1844:39).

THE FOREIGN COMMUNITY

Even if our travellers were not invited to stay in the homes of consuls or
merchants, they had letters of reference to such people. In one way or
another, most soon took up contact with someone among the foreigners
in Veracruz. They were especially sensitive to the living conditions—
and the fortunes—of their countrymen. These, in turn, felt isolated and

welcomed new arrivals, even if not personally known to them. What a thrill to hear news from the homeland; how reassuring to sing old songs in one's own language on a warm evening deep inside a tropical town (Koppe 1835:176). One can quickly identify several national subgroups, but also a certain unity among Europeans. They were all more or less convinced that whatever refinement the town might boast issued from them—flickering signal-fires in a night of barbarism (Buschmann in Thümmel 1848:93–4). This foreign minority and the several other elements of the population which the travellers noticed on arrival or subsequent walks through the streets are what we are given to see of the population of the town; the majority is beyond our purview.

There is only one numerical approximation of the foreign population of Veracruz in the literature basic to this analysis. It comes from Koppe's review of information available to him on the demography of the state in 1830 and 1831 (Koppe 1837, 1:153), and is shown in Table 2.

TABLE 2: Inhabitants of the Canton of Veracruz* by caste and origin, 1831

Mexicans (Creoles, Indians, a few negroes, and many mixtures of the three)	23,799
Cubans	114
Peninsular Spanish	294
Other foreigners	349
	24,556

*Including the port, plus Alvarado, Medellín, and the rural surroundings of all three.

From the same source comes what seems the most reliable count of the population of the city in the first few post-Independence decades: 6,828 for the town proper, plus another 687 in adjoining villages for a total of 7,515, of which 3,636 were male and 3,879 female (Koppe 1837, 1:156). (Scattered comments by various authors indicate that among the "other" foreigners there was a decided sexual disproportion in the other direction.) If one considers that almost all the 349 "other" foreigners in the canton will have lived in the town itself, it is clear that when we follow the German, English and French travellers, the vast majority of our group, into the foreign colony, we are shown the living conditions and concerns of about 5% of the population of the town.

In a recent analysis of the German presence in nineteenth-century Mexico some numbers are given on their community in Veracruz (von Mentz de Boege et al. 1982:149–59). But like so many other tabulations out of this epoch, they are isolated windows on reality, difficult to put into context. As many as one-quarter of Koppe's foreign population may have been German. Most of those were people of substance, that is merchants. By the end of our period there would also be hacendados, professionals, employees of the railway, artisans with means, as well as some "poor" Germans too, such as ex-soldiers from Maximilian's forces and artisans without means.

Trade was by far the most important economic activity of Veracruz. Its most lucrative aspect was the wholesaling of imported goods, which was dominated by the resident foreigners, and they were single-minded: "Mammon is the sole god of the city which is called after the symbol of our faith; and here the bones of worshippers whiten the sands" (Latrobe 1836:299).

Robertson describes the establishment of one wholesale trader, "a principal personage in the port" and probably an Englishman (1853:1,252): "It contains, under one roof, counting houses, warehouses, vaults, wine cellars, stables, coach houses, kitchen, servants' offices and rooms, drawing rooms (three or four), anti-chambers, dining rooms, breakfast room, bed-rooms, corridors, balconies, all including the open courts, on a gigantic scale and, of course, occupying, as a whole, an immense area" (Robertson 1853:1,256).

Robertson was seeing an example of an *almacén*, which represented an important early phase of European commerce in Mexico (Pferde-kamp 1958:35–67). Those seen in Veracruz were often subsidiaries of central houses in Mexico City, from which the trade of the country was dominated. An *almacén* typically required for its operation one or two sailing shiploads per year of a variety of manufactured articles, received on consignment and in turn sold on credit that extended from one ship's arrival to the next. News of such arrivals spread fast. Store owners from the interior soon converged on Veracruz with their mule trains. The importer hardly needed to leave his office. These were re-enactments, on a reduced scale, of the excitement of the fairs that took place upon the arrival of fleets in colonial times, particularly in Jalapa.

In the 1820s English houses dominated the importation of all manner of consumer goods. Germans entered the competition in 1823 with a joint stock company, the Rheinisch-Westindische Kompanie. In the course of the next several decades this company folded, but newer independent German firms expanded the German hold on the market to such an extent that *almacén* became synonymous with *alemán*. Between 1850 and 1860 they were at their height. In the meantime the English houses went more and more into banking.

During the 1830s the French entered the picture, first as retailers of imported articles, getting their stocks from German *almacenes*. Spaniards were entering in a similar way. Together they were overwhelming native retailers. They were immigrants of modest means, some of them farmers' sons, trying to find their feet. Eventually the French were able to give the *almacenes* competition in their own field. Moreover, improvements in communication and transportation loosened the stranglehold of the large houses on import trade generally, rather as modernization of the road network during the last few decades has lessened the hold of powerful middlemen in remoter areas of Veracruz. The Americans participated in the ramification of import trading by entering as travelling salesmen. One such salesman, the dreadfully disappointed Pattie, passes through these pages. They were not prominent along the road into Mexico, but American entrepreneurs of other sorts were, as will be seen.

The imported goods that were traded wholesale make evocative lists. These are available for 1828 (Wappäus 1863:99), 1831 (Koppe 1837, II:239) and 1851 (Wappäus 1863:100–1). Textile manufactures of various forms were in first place throughout the period: articles made of linen, silk, cotton, and wool. Paper, glass products of many kinds, porcelain, furniture, as well as all the iron and steel requirements of the country, had to be brought in via Veracruz. Alcoholic beverages were always a part of the trade, as well as cacao and tea. The lists continue with wax, salted fish, medicines, instruments—in short, the whole intriguing, fragrant inventory of goods Mexico needed but could not produce for herself.

Through Veracruz in the other direction, it might be remarked in passing, went a less diverse list of goods, but with total values that often surpassed those of the imports. In the early post-Independence years export was limited to precious metals and cochineal; later, lumber, hides, exotic plant products like vanilla and pepper, and a certain purgative from Jalapa were also part of the traffic.

Many of the authors referred to in this book had the responsibility, or took it upon themselves anyway, to advise the aspiring trader at home on what he would face in Mexico and how he might proceed. Bullock remarked in general terms on the importance of understanding existing demand, and the need, moreover, to prepare for the fulfilment of the demand that the introduction of European fashions would generate (Bullock 1825, II:224). Koppe prowled the commercial houses of Veracruz and the capital at various times over a twenty-one-month period in 1830 and 1831, remarking on stocks and the state of trade, on which countries were doing best in regard to which clothes, giving advice to his German countrymen as he went, much of it sounding like very good sense. He cautioned prospective traders to be very careful about Mexican wants and prejudices, for example, their taste for certain bright colours. He ruefully noted a fundamental predilection: French cosmetics and articles of high fashion were in great demand.

Trading through Veracruz, obviously profitable enough throughout our period in order to flourish in spite of everything, was beset with formidable difficulties. First among them were the duties on goods coming into the country:

> Eight and a half percent *ad valorem* is paid on all cargoes from Europe, at the Spanish castle of St. Juan de Ulúa, and twenty-seven and a half to the town. This too is on their own arbitrary valuation, and is often three times the original cost. One dollar each package is charged for the hospital, and four and a half dollars per ton for water; eight dollars for the captain of the port; and thirty-two dollars each trip for the use of large boats for landing the cargo: to these may be added the expense of porterage to the custom-house, and afterwards to the stores of the merchants; for even this is expensive, as labour of all kinds is here excessively high, and the insolence of the negro porters is intolerable. The above are the principal charges at the port, to which the removal of goods to Mexico adds much, as they pay an additional duty of about twelve per cent on their arrival, and the carriage of every horse or mule-load, from two to three hundred weight, is from eighteen to twenty-two dollars. The goods sold in Mexico pay again another duty on being removed to the provinces; but if they are intended, when landed, to be removed direct to the cities beyond Mexico, an arrangement can be made at the port custom-house which will save some of the expenses. (Bullock 1825, II:220–2)

In the early 1850s a merchant in Veracruz outlined some additional problems to the French traveller, Valois (Valois 1861:46–7). Goods moving along the roads were often lost to bandits. Peremptory levies were often made on merchants, for war costs or for local government expenses. Valois felt the Mexicans were not entrepreneurial enough to get the things they needed from abroad, and only too ready to harrass and slander those who did.

As might be imagined, travellers found prices high in Veracruz, especially for imported articles. One had to be a Rothschild to have a bottle of Bordeaux with one's dinner (Valois 1861:91). But costs of local labour and transport were also thought to be exorbitant, as was the cost of food. Vegetables were much more expensive in this port than in Havana (Valois 1861:45).

Of course, the foreign merchants of Veracruz were not hapless in the face of the insecurity around them and the always desperate need of the politicians for money. Indeed, the foreign and native merchants of the port were seen as politically manipulative, and as having profited from the decades of disorders (Fernando Ramirez in Pasquel 1979:331–3;

Kollonitz 1867:70). They had found ways of circumventing tariffs and levies, while keeping prices high. There is a documented case of a prominent German merchant who was arrested for having traded in contraband (von Mentz de Boege et al. 1982:151).

For this and other reasons, some travellers were not too taken with their fellow countrymen abroad. They saw them drinking too much, shouting obscenities in the cafes or in other ways letting their end down. Various travellers noticed sallow foreign faces; Valois thought his fellow Frenchmen looked dead, having come to earth to die a second time (1861:42-3). These had to be avaricious people; no one without that motivation could stand to live in such a dismal town (Heller 1853:35-6). If they did not die of yellow fever they might well become rich, but they would "return with shattered constitutions" (Mayer 1844:3).

The merchants, as Valois' report on his various conversations in Veracruz indicates, thought of themselves as salt of the earth. Mostly they kept ledgers, not diaries, and they went to Medellín on the weekend. This was a village just south of Veracruz, set in the wooded surroundings. Koppe happened on it in the spring of 1832 and found both native and foreign members of the port's high society disporting themselves (Koppe 1837, II:315-17). Some forty years later, Ratzel found it to be an even livelier resort, now with its own rail connection to Veracruz (Ratzel 1878/1969:158-9).

In Medellín one could go swimming, drink, gourmandize, gamble, dance, and in other ways compensate for the rigours of life on this coast. Many of the foreigners working their way up in the various trading firms were single men. Some of the affluent Veracruzan families had summer homes in the village and *haciendas* on the relatively humid or irrigable soils in the vicinity. The surroundings of the town offered incoming travellers a first look at tropical luxuriance. Valois described the greenery and the wildlife seen on such an excursion. But he and others like him soon gravitated to some shaded establishment along the river for entertainment where, like as not, the veterans of life in Veracruz were already foregathered (Valois 1861:64-5). Medellín is a rather lethargic village now; it has long since lost its recreational importance, although some of the imposing old summer houses are still there. Agrarian reform has affected its large private properties, and major roads through the central part of the state have bypassed it.

Several travellers commented on high society as they had observed it in town. Valois noted that it was narrowly based. Once the merchants had made their fortunes they were likely to move to Jalapa (Valois 1861:45-50). Payno, the well-connected Mexican traveller, elaborated on the same point (Tavera Alfaro 1964:111-13). The havoc caused by the struggle of Independence, the expulsion of the Spaniards, and what he perceived as a deterioration of commerce in Veracruz, had reduced the acclimatised population. Of more immediate importance to him was the fact that he could have a fine time in Veracruz with his gentlemen friends, but there was a decided scarcity of girls. They were more susceptible than males, he maintained, to the rigours of this climate and especially to yellow fever. The ladies of better families were thus often sent to Jalapa or elsewhere while the men stayed behind—a point made repeatedly in this literature. The actual numbers involved cannot have been very large. The one available count of males and females in the entire population of the city shows an imbalance in the other direction (Koppe 1837,1:156).

Payno, a native of the capital, also commented on an idea that persists in Veracruz to this day: Veracruzanos have held a disproportionately high number of leading positions in Mexican government and economy. As though he wished to show himself unsusceptible to the uplander's old contempt for the lowlander, he professed to have noted the special ardour, initiative, frankness, amiability, and indeed a certain nobility, among the better people of Veracruz.

The foreign observers do not seem to have noticed much of this. Several authors did remark that frequent contact with foreigners had brought a greater knowledge of the world to Veracruzanos than one would find among the people of other cities of the republic, even Mexico City. They also seemed less fanatical in matters of religion, more open and accepting (de Fossey 1964:252; Biart 1959:21). Valois, the well-connected French traveller, was invited to a ball at the house of one of the most important foreign merchants in Veracruz, at which the Mexican high society of the town as well as foreigners were present (Valois 1861:45–50). He found the house unimpressive, without the luxuries one might expect—thus echoing judgments Madame Calderón de la Barca and other foreign observers had made before him. He was struck with the affectations of the Mexican guests. The ladies and gentlemen he saw had abandoned their own customs and natural graces and spared no effort to appear Europeanized, some in French and others in English styles. Moreover, Valois felt that they had adopted from European customs only the ridiculous. He was appalled to hear Mexicans speaking Spanish with a French accent. Flighty, tangential chatter, flowery turns of phrase, and plays on words were considered very chic. He suspected they must have learnt it all from French travelling salesmen.

Valois was especially critical of the Mexican ladies. They chattered, played with their fans, listened to an insipid instrument called the guitar, and smoked, which scandalized foreign visitors one and all. The ladies were pretty enough: ravishing little feet, soft eyes, deeply red mouths, silky black hair, and admirable figures. But there was something missing, which Valois could not put his finger on, except to call it the radiance of inner beauty.

IN THE STREETS

Biart observed that Veracruz was French in its customs, Spanish in its form, but actually populated by *mulatos* (Biart 1959:17). These were the vast majority of the population. Foreign visitors seldom socialized with them, or even conversed with them at any length. For some insights into their lives and some further indications of how the people of this town were seen we need to follow the visitors on their brief walks about the town, and short excursions into the immediate vicinity, during their period of "detention." This material has little factual depth, but many perceptual facets.

The wide range of skin tones of the passers-by was itself remarkable, everything from black to a dirty white. This was novel, but hardly attractive (Becher 1834:38). The old prejudice against people of mixed racial background emerges frequently in this literature.

One blesses Koppe here, the careful observer, upbeat both at the beginning of his Mexican journey and at the end, deserving of more frequent consultation in the study of nineteenth-century Mexico. The streets of his Veracruz are crowded with variety too, but the many skin tones are painted with a fresh palette, not the grimy tones other authors often gave the racial mixture of the port. The costumes are a bright array; no mention of half-nakedness. He recognizes that the vegetation in and around the town is not that abundant, but this is no desert. Every palm, every mimosa, every cactus is a wonder to someone from the temperate zone. The streetcleaner's crude broom is a bundle of curious species. The offerings of the fishmarket and the very vultures on the rooftops are sights, he feels, that would be the envy of any European professor of natural history (Koppe 1835:168).

The movement and noise of the streets emerges from many accounts. It peaked twice each day, a basic fact about any Spanish-American city. Biart awoke in his hotel near the open square behind the wharf to the noises of a town already at work. Goods were being wrestled about, carts thumped over the cobblestones, vendors and muleteers shouted their strange shouts, and church bells pealed (Biart 1959:18). Others heard the bells at the necks of donkeys carrying the casks of the water-sellers. Veracruz relied mainly on cisterns for its water supply. There was water just below the surface, but it was brackish. Hence carrying in fresh drinking water from some source beyond the sandy region was good business. Milk was sold door to door from metal cans on the backs of horses or donkeys. And there was an additional sound: the clank of the chains on the feet of the prisoners who had to clean the streets.

Payno observed the waterborne equivalent of this bustle in the harbour on a day when there were no important disembarkations, only the daily along-shore traffic (Tavera Alfaro 1964:102). He saw all sorts of small craft and noted vegetables, meat and even barrels of water in transit. Salt and other merchandise was coming from Alvarado and had perhaps originated somewhere in the Papaloapan basin.

Biart was in the market just after midday. He had been deafened by a hundred varied shouts and then noticed how the noise level abated. Buyers and sellers eventually disappeared and the city was left deserted. He returned to his hotel himself, tired and wet with perspiration.

Müller watched tradesmen work out on the sidewalks in front of their shops once the sun had descended (Müller 1864:199). Families were seated around the doors, conversing in low tones, and taking their reward for another day spent in a tropical town.

Koppe obtained some official information on occupations in Veracruz (see Table 3). Deducing from this what applied to Veracruz and drawing in stray comments from Koppe and others allows us to summarize the main economic activities of the town. Service was undoubtedly in first place, including the work of lightermen, day labourers, and domestics. Small manufacturing establishments, such as weaving mills, rum distilleries, and brickworks, as well as a large number of craftsmen, produced some of the basic consumer goods, but the greater part still had to be imported. There was no shipbuilding and only minimal repairing. Commerce at various levels accounted for the third main group and government administration the fourth.

Loquacious Mr. Robertson was on hand one Wednesday in February at the arrival of the *conducta*, a periodic transport of precious metals from Mexico City to the coast (Robertson 1853:1, 248-9). It was one of the largest ever, made up of a train of sixty wagons, managed by mounted muleteers, and escorted by a military force. It evidently carried specie and bullion worth one and a quarter million dollars, all neatly packaged in matting, obviously a magnificent sight. The treasure was taken first into the warehouse of an English company and then carted from there to the wharf, which took two entire days. This was the aspect of Mexico, of course, that had fascinated Englishmen for centuries. An equally valuable cargo was also being carted out to the same ship: over 500 bales of cochineal, each weighing 200 pounds. But this was not as intriguing and rated only a few lines.

There are descriptions of visits to markets in 1832, 1843, 1846, and 1851 (Becher 1834:37-8; Payno in Tavera Alfaro 1964:107; Biart 1959:19; Valois 1861:42). The market was abundantly stocked each time

TABLE 3: Occupations of grown men, excluding armed forces, in the cantón of Veracruz (including the town itself, plus the villages of Alvarado and Medellín, 21 haciendas, and 600 ranchos), 1830 (Koppe 1837:I,174-5)

Priests	21
Schoolteachers	10
Government officials	237
Merchants, wholesale and retail	767
Doctors	13
Druggists	11
Lawyers	7
Factory owners	32
Notaries	3
Craftsmen	861
Farmers (and herders?)	3,759
Fishermen	19
Lightermen	114
Day labourers	3,222
Domestics	872
	9,948

Out of a total population for the cantón of 24,556, or an "economically active" sector of 40.5%

with grains, fruits, vegetables, fish, and meats. Payno added that this should not be surprising, since even though the town was surrounded by sand, it could be supplied from Medellín, Boca del Río, and Alvarado. Agriculture is also strongly represented in Koppe's table of occupations. Nevertheless, travellers probably read too much abundance out of this, as will be developed.

ENTERTAINMENTS

By way of additional diversions, the traveller had several options. One of these was the occasional bullfight; the ring is visible in Figure 6. Valois went to such a fight and saw all the horrors he had read about: horses pulling out their bowels, bulls being tortured, and men wounded. The beautiful women applauded like cannibals. Trembling, sated of blood, and with the screams of the crowd in his ears, he left. Surely people who needed this sort of diversion were condemned by God (Valois 1861:69-70).

There was entertainment of quite another sort available on the waterfront during a *norte*. The watergate provided a proscenium; if one was watching from an *azotea* the whole horizon became a stage. The sky and the water darkened rapidly, the wind rose and small boats scattered from the wharf like a flock of gulls (Payno in Tavera Alfaro 1964:103). In a short time communication between ships and shore would be lost, and it might be necessary to put out to sea in order to ride out the storm. For those on shore, it was a fine show of waves and a wild sky.

The wind brought inconvenience, but it cleared the air of any miasma lingering from the rainy season and, of course, one was in no danger of shipwreck on dry land. Residents shuttered and barred their doors and windows in order to keep household articles from flying about and to keep out at least some of the sand that was blowing in from the dunes. Pedestrians venturing through the streets had to flatten themselves against the walls. Caught by a gust, they might well be made to retreat in huge backward steps over territory laboriously conquered (de Fossey in Glantz 1964:251). And, one traveller noted: "the washerwoman refused to take the clothes, unless I would release her from all responsibility if a Norther should carry them away, whereupon assuming the hazard, on the following day, upon short notice, the winds came, and scattered my clothing like kites in the air, some to the country and some to the sea, and some perhaps to needy Mexicans" (Gilliam 1847:24).

A serious *norte*, with many ships at anchor between the fort and the town, was high drama. Such a storm occurred in 1852 and was described by the American traveller, Wilson. It is one of his best passages. There was a crowd watching the scene. Crewmen on the ships were throwing out all extra anchors, but a coral bottom is only a poor holding ground. One by one the ships began to drag off to the right, southward, toward the reefs and the shore itself. They were dashed to pieces, and then began "the individual struggle of each separate mariner, made in the very sight of those who could render no assistance but must stand idle spectators" (Wilson 1855:21-2).

Something else was happening in the harbour during this storm-diagnostic of developments in transportation technology. A small steamer on a single anchor was able to ride out the storm by keeping its paddlewheels in steady forward motion. The clouds, wind, waves, and shipwrecks, and the steam engine in the midst of it all, invoke J.M.W. Turner's romantic paintings, especially those out of the 1840s, such as *A Stormy Shore* or *Fire at Sea* (Clark 1976:181–97). Wind and water had been transformed into colour and swirl.

Ferry described the spectacle of a *norte* on the night of his ride through the dunes southwards to Boca del Río, where he was to witness wreckers at work: "Forced sometimes to turn my back to avoid the clouds of drifting sand, I now and then had a glimpse of the town that I repented of having left. At regular intervals, the light-house of San Juan de Ulloa blazed up in all the beauty of its rendering light, sometimes gleaming on Vera Cruz shrouded in darkness and then on the roadstead white with foam. For a moment I discerned the ships at their anchors pitching up and down on the broken swell, and almost driving on each other: the light soon turned and all was dark" (Ferry 1856:319-20).

There were various places in town, besides the balcony from which Koppe had looked down, and to which Valois had looked up, where travellers could get a good look at the women of Veracruz. One was at the café under the arcades opposite the cathedral. In the morning the ladies could be seen coming out of mass, "bonnetless women of low and high degree are returning to their homes; some wearing mantillas of satin, black and shining as their raven hair, which are pinned by a jeweled pin upon the top of their heads; others, more modern in their tastes, sport India shawls; while the common class still cling to the 'rebosa' which they so ingeniously twirl around their heads and chests as to include in its narrow folds their arms, and all above the waist except the face" (Wilson 1855:16).

One could also scan the crowd at the theatre, a decidedly third-rate affair. Payno, from his urbane, upland perspective, describes one particularly interesting group of girls that he noticed there, the *jarochas*. They represented a sub-culture associated mainly with the villages and the countryside of Central Veracruz. At the theatre they were the charming provincials in costume set off from the Europeanized townspeople. They did not sit with the better families, but they were very clean and very gracious; they had lovely teeth, and of course, black expressive eyes, as well as fine facial features. There was something Andalusian about them. Usually, our observer points out, they were *morenas*, the light, ambiguously brown-skinned girls who had always fascinated gentlemen of Spanish background.

The best opportunities came during the ritualized early evening walk, the *paseo*. People walked up and down the one wharf and around the central plaza; evidently the latter was something of a preserve of the fashionable set. Most agreeable, it seems, was a walk out along the *alameda*, a shaded avenue beyond the town walls. It began outside the Merced gate, about where Independencia meets O'Campo today. On the Casimiro oblique aerial view (Figure 6) it is marked by a double line of trees, which coincides with the modern Avenida Salvador Diaz Mirón. It was no grand avenue. However, the Veracruzanos appreciated it a great deal because of the relative scarcity of flowers and trees in their town, certainly when compared with Jalapa, and also because of the narrowness of the patios in many houses (Estrada y Zenea 1872:66).

Here was a place to which the ordinary woman, circumscribed both by the infrequency of cultural events in this town and the rules of propriety, might escape the prison of her home and take the air.

Müller, the soulful naturalist, took every opportunity during his stay in the town to go out on the *alameda* (Müller 1864:206). He was stunned by the way the Veracruzanas made luminous jewelry out of fireflies. He had already been captivated by pretty feet and black flashing eyes, but this was something more. The firefly is called a *cucujo*, or *Pyrophorus clarius*, as he felt obliged to inform the reader: it was captured out in the country by the Indians, brought to the town market and then fed by the ladies for use as needed. It could be skewered without being killed and used to grace a hairpin, or it was put into diaphanous little sacks that were sewn as bordering ornaments on a dress.

THE HARD-PRESSED TOWN

Müller's *alameda*, Koppe's crowded streets, and here and there throughout the commentary of our company of travellers a scattering of observations born out of fascination, all these belong to an exotic Veracruz, noticed by those who seem to have been able to put aside the threat of yellow fever. It counter-balanced somewhat the macabre town. But there was another town in the accounts as well, mostly obscured by the exotic and the lurid. It emerges from isolated observations in accounts the gist of which was otherwise; it predominates in the portrait of the Veracruz of these same decades drawn by Trens, a Mexican historian, some of whose deductions, drawn mostly, it would seem, from native sources, may be brought in here to corroborate those fleeting comments of the foreign travellers, but also to set the tendencies of their imagery into relief.

Valois, always highly articulate and disparaging of most things Mexican, went out along the *alameda* (1861:60–1). He was not impressed by the miserable acacias that lined the avenue. The slaughterhouse, noted in the Castro print (Figure 6), and a church ruined by American cannonballs four years before, depressed him further. Along the avenue he met young women carrying dead infants on their heads, the cemetery lay just off the *alameda*. The children had been arranged in baskets with flowers and ribbons. They would be buried without the assistance of a priest. These were businessmen, he added, who sold prayers as one might sell wine or cloth, and did not pray on credit. Someone had told Valois, and since he did have good contacts in Veracruz the information probably had some validity, that infanticide was frequent, an indication of very difficult living conditions among at least some sectors of the population of this town.

Decay was apparent to visitors throughout this *época negra*; there are references to vacant houses within the town walls. Their roofs had fallen and their patios were full of a tangle of untended vegetation. Visitors who ventured on excursions beyond the walls found extremely dilapidated housing in the nearby villages as well, plus poor hygiene and wretched poverty. Madame Calderón de la Barca remarked on the dilapidation after she and a distinguished party of companions had been out for a walk in the environs of Veracruz: "All looks as if the prophet Jeremiah has passed through the city denouncing woe to all the dwellers thereof" (Calderón de la Barca, Fisher ed. 1966:59). She was seeing this in 1839. Müller observed something similar in 1856. He found the crumbling houses picturesque, but wondered why they were not being reconstructed (Müller 1864:199). There seems to have been very little new building in Veracruz throughout our period.

One can put together a table of the population of Veracruz over the first half of the nineteenth century discussed here, relying on those figures cited from more or less official sources by authors otherwise careful in their treatment of this sort of data:

TABLE 4: Population of town of Veracruz

Year	Town	Source
1803	16,000	Humboldt
1831	6,828	Koppe
1844	7,000	Trens
1866	11,000	de Emperán

It is immediately apparent that Veracruz was much reduced by the concomitants of Independence and was recovering slowly, at a generalized rate of perhaps 1.5 per cent per annum. Parallel figures are not available for the surrounding region, but there are several indications that the population of the state was growing even more slowly in the decades after Independence.

The most difficult times during the decades covered by the travellers' accounts came during the sequence of blockades and bombardments already outlined in Chapter 1. Political unrest disrupted transportation, which impeded commerce. The town repeatedly felt itself isolated from the rest of the country, made to bear the brunt of the country's unfortunate foreign relations without much help from the central government (Trens 1950:64–110).

During the struggles for the country's independence only occasional shipments of food and merchandise reached the town. There was no income and no help from the rest of the country in the town's defence. By 1813, what with commerce paralysed, property in Veracruz had evidently lost two-thirds of its value, industry was in ruins, and the agriculture in the surroundings was prostrate as well.

In the period between 1821 and 1825, when the fort was still in Spanish hands and the town already freed, trade was again impeded; in fact, foreign trade was carried on temporarily from Alvarado. The new trading opportunities consequent on Independence could not be fully grasped. The bombardment between the town and the fort from 1821 to 1825 obliged a great many citizens to flee into the surrounding country, where they lived under very difficult circumstances, often camping out (Calderón de la Barca, Fisher ed. 1966:612; Trens 1950:78).

Ward arrived in Veracruz after the town had been thus abandoned. Only a scanty force of defenders remained; visitors passed ruined churches and houses that had been riddled with shot. Their own footfalls made the only sound, and vultures were the only sign of life. Then, suddenly, the party of Englishmen came on the houses where the defending General Victoria and his officers were loudly celebrating his saint's day. In a surrealistic switch a tremendous welcome ensued, since, of course, the establishment of a diplomatic relationship with England was very important to the man who shortly was to become the president of the country. When Ward's party was ready to set out for the interior they found there were other people who had remained behind, "dregs of the population" from which they had to find muleteers and coachmen: "They were almost all blacks, or descendants of blacks, with a mixture of Indian blood, and seemed either never to have known the restraints of civilization, or, at all events, to have lost sight of them amidst the wild scenes of the Revolution" (Ward 1828, II:174–8).

Similar dislocations followed, as during Santa Anna's uprising from a Veracruzan base in 1832. Tudor arrived just before the conflict:

The country was on the eve of civil war. Numbers of the fair sex had already left the place, the rest were preparing to follow their example—commerce was entirely suspended—the merchants were fortifying their houses with barricades of cotton bales, with which they were blocking up their doors and passages, and the lower range of windows—troops, both horse and foot, were flocking into town—and everything gave "dreadful note of preparation." The government forces were lying about three leagues off, at a small village called Santa Fé, from which the inhabitants were expecting an immediate attack (Tudor 1834:156–7)

During the French blockade of the port in 1838 the inhabitants were obliged to flee once again—a three-month exile. Maissin observed "a sort of Pompeii, not in ruins, but equally dead. The houses were shut up, the streets so little used that weeds sprang up in them. An unbroken silence reigned throughout. Only a small patrol, always ready to flee to the interior, kept watch on the sleeping city and protected it against vandals" (Maissin 1839/1961:89).

During the siege and bombardment of Veracruz by the Americans in March 1847, many buildings came down on their occupants and others were burned out. Because the town was mostly one of masonry, there was no general conflagration. The food ran out and there were no safe places to put the wounded. Projectiles crashed through the domes of several churches where the sick had been laid out. The well-to-do, who might have been of great assistance, had left the town before the siege began (Trens 1955:94–107). The bombardment was horrible to watch from the outside. Shells arched through the sky by day and by night. Many fires were burning. The yells and screams of the victims could be heard by the attackers (McCall in McWhiney and McWhiney 1969:12).

The unrest, the scarcities, and the hardships of other sorts during the post-Independence decades induced many merchants, large and small, as well as craftsmen, and indeed anyone who despaired of making a living in the port, to pick up what they could and move inland. Undoubtedly, even some of those who thought only temporarily to evacuate their premises will have stayed away. The out-migration was to some extent countered by the influx of foreigners and Mexicans capitalizing on the new trade opportunities, but this was a minor factor, demographically. The out-migration, more than any other factor, led to the tremendous drop in population between Humboldt's figure of 16,000 for 1803 and the 6,828 counted in Veracruz in 1831. It can hardly be attributed to one single outbreak of cholera, nor indeed to the ravages of disease alone (Boyer 1972:155). Yellow fever, often cited as the main killer in Veracruz, was a relatively minor factor in the mortality among the native population. Other diseases carried away as many or more. Diseases in general, of course, would have contributed substantially, and perhaps infanticide as well, to the very low growth rate and the widespread misery.

As one might expect, foreign observers repeatedly judged living conditions in the town by what they saw of its commerce; there are some problems in interpreting what they describe. Often they remarked on the well-stocked Veracruzan market, and they deduced abundance from this. Trens in his history of Veracruz repeatedly refers to scarcities in all basic necessities (1955). The produce was obtainable in neighbouring villages, such as Medellín and Alvarado. However, transport from there into town could easily be disrupted and it is likely that supplies varied. It is quite possible that the travellers' samplings missed the occasions of

scarcity. More likely, they were not able to assess correctly the significance of the mounds of fruit and vegetables, the fish and the meat. It all
appeared very exotic. It may not have been accessible to most of the
urban population. The travellers repeatedly refer to the high prices of
both fresh produce and locally crafted articles, as well as services such
as cartage, and certainly imported goods (e.g., Thümmel 1848:94–5).
Ships tried not to take on stores in Veracruz. In any case, an intriguing
disparity remains. Something similar will be noted with respect to the
reports by travellers on the Jalapa market and the laments of its historian, Rivera Cambas.

Foreign trade in and out of Veracruz, although frequently disrupted,
is shown to have fared rather well during the *época negra*. It was noted
that as early as 1824 the total value of Mexican exports and imports had
reached the average annual levels that obtained shortly before Independence (Ward 1828, 1:436). Subsequent analysis indicates that
between 1825 and 1851 imports increased 4.6 times and exports even
more (Herrera Canales 1977:114). Although some post-Independence
trade flowed through other Gulf ports besides Veracruz, including Alvarado during the periods of closure, surely the old "spigot of New Spain"
recaptured at least its pre-Independence importance.

However, foreign trade was largely a through trade, without too much
direct effect on the economy of ordinary people in Veracruz, except
those employed in the *almacenes* and the transport services. During
closure, of course, this employment, too, would be affected. The permanent shift of an *almacén* to Jalapa or Mexico, which was evidently quite
frequent during our period, would also aggravate the local economy. By
mid-century the *almacén* had had its heyday in any case.

It was undoubtedly disturbing for the officials of Veracruz that the
remaining merchants were avoiding tariffs as much as possible. It is
mentioned in various places, but understandably not well substantiated
(e.g., Kollonitz 1867:70), that merchants were actually exploiting the
disorder they deplored. Contraband remained the constant of foreign
trade along this coast after Independence that it had been before.
Valois was told of aggressive town officials who virtually waylaid foreign
merchants in order to extract contributions from them (Valois
1861:46–7). Clearly, the town treasury was not getting what it might
have. More important for the general economic health of the town than
all of this may have been the impact of disturbances and isolation on
local retail commerce.

THE VARIOUS TOWNS

Reading the foreigners' descriptions of Veracruz, pursuing one after
another of the relevant passages in a literature that spans half a century

does indeed allow one gradually to sense the walls and feel the pulse of the town, to gain some understanding of the provincial "urban scene" of an epoch (Arreola 1982:2). However, the purpose here has been not so much to deduce what was actually there as to analyse what was seen to be there, and in particular to interpret the judgments on the port as a beginning in earnest of the characterization of a tropical lowland.

Veracruz, for most of the observers, exemplified the worst aspects of tropical nature and tropical man. It was located on a barren beach, was uncongenial in form, occupied by canaille and edged by a ghoulish glow. One expected to be compensated by the greenery beyond the dunes. Some noticed the unfortunate town, especially if they happened to pass through it after it had been evacuated or bombarded. The exotic town was taken in by the sensitive and curious, left alone by others.

It amazed Madame Calderón de la Barca that anyone could come to like this place. Behind the imagery already traced, behind this play of shadows, there was still another Veracruz: the preferred place, the home town. That Veracruz is almost completely beyond our purview; one would need to search for it in Veracruzan poetry and graphic art. People who came to Veracruz to live, Madame maintained, "even foreigners, almost invariably become attached to it; and as for those born here, they are the truest of patriots, holding up Veracruz as superior to all other parts of the world" (Calderón de la Barca, Fisher ed. 1966:59).

CHAPTER FOUR

Guardian Dragon

While waiting in one of the hotels of Veracruz or walking its streets the travellers seemed to be feeling their brows every so often in order to find out if there was not already a sign of fever. In the observations and reflections entrusted to their journals during this detention, more than at any other time in their journey inland, they wrestled with the phantasmagoria of yellow fever, *el vómito negro*.

From whatever information was available to them they assessed the likelihood of their infection. Both the yearly toll and the credibility of the reporting vary tremendously. During May of 1834 Latrobe heard of forty deaths a day. That year the disease had persisted throughout the winter months; as usual it was expected to get worse from August to October (Latrobe 1836:299). The one available set of plausible figures on fatalities in Veracruzan hospitals attributes 155 out of 1,017 deaths in 1841 to yellow fever (Mayer 1844:7). In 1845 Heller noted that the disease yearly claimed an average of two-thirds of all newly arrived Europeans (Heller 1853:36). This will have interested his German readership, many of whom were considering emigration! In 1848 Robertson was moved to record, rather unctuously, that "the course and nature of the disease being well understood, it never, *with common care*, now proves fatal" (1853:260). No one seems to have agreed with him in print. Wappäus, the German geographer, compiling his sourcebook on mid-century Mexico, ventured a generalization: about half the strangers that arrived in Veracruz during the summer months from abroad or from the uplands died of the disease (Wappäus 1863:153). That sort of risk sharpens sensibilities.

From colonial times until the beginning of the twentieth century the port more than the adjoining lowlands was associated with the disease, and of course no one yet knew how urban a malady it really was, but when it was actually discussed the danger was perceived to come from somewhere just beyond the walls of the town. In attempting to come to terms with the disease, therefore, the travellers were continuing their reflections on the Mexican landscape begun with the first glimpse of Mount Orizaba, focused now on the dunes and more particularly the swamps among and beyond them, a truly dreadful aspect of lowland luxuriance. Out of the resulting journal entries comes a great deal of the lore of the disease, not only prevalent ideas of the day about its origin and environmental limits, but also about its prevention and treatment. And there are many indications of the effects of the fear of it. All this can be set into strong relief with a consideration of what is now known. This gives one the sensation of being present at a parlour game of long ago, watching blindfolded players trying to pin the tail on a donkey, except there is little amusement.

THE DISEASE AND ITS "CONQUEST"

Yellow fever is a viral infection transmitted within settlements by the *Aedes aegypti* mosquito and among mammalian hosts in the tree tops of the tropical forest by various other species of mosquitoes. Infection is followed by several days of incubation and then a series of severe symptoms: headache, backache, fever, nausea, and vomiting. There are two classic symptoms: hemorrhaging into mucous membranes blackens the vomit and gives the disease its common Spanish name, *vómito negro*; bile pigment from destroyed liver cells yellows the skin and eyes, which is reflected in the English name. There is a great range in the severity of the disease, from an infection with hardly any symptoms, sustained in childhood, to a severe case that kills an adult in six to nine days. Often it is difficult to distinguish from related infections, such as hepatitis. It can be prevented and ameliorated, but not cured.

Insects had been suspected as disease carriers in various ancient Near Eastern cultures (Leake in Smith 1951:32). Around 1800, while Humboldt was travelling in the Americas, an Irish physician, John Crawford, made an explicit connection between yellow fever and insects. This was elaborated in 1849 by an American doctor, Josiah C. Knott. He was forceful in his arguments against a theory of disease that indicted exhalations, but these ideas found little resonance. In the early 1850s, a Venezuelan doctor, Beauperthuy, suggested that mosquitoes were responsible for the transfer of the yellow fever poison from marshes to human tissues. The germ theory was put forward by Joseph Lister, the British surgeon, in the late 1860s.

These intimations were taken an important step further by a Cuban doctor, Carlos J. Finlay. He discredited miasmas categorically, proposed *Aedes aegypti* as the vector, and declared that it carried the disease from person to person. Unfortunately, his laboratory techniques were faulty, his facts not quite sufficient, and some preconceived ideas too tenacious. He became known as a genial and lovable crank, and missed the distinction of establishing the truth of his own hypothesis. This went to Walter Reed in 1900. Soon after, the work of eradicating the mosquito could begin. The key to this eradication was the realization that the disease was predominantly urban. *Aedes aegypti* likes to breed in places like the containers of drinking water in the home, the puddles of rain or sewage in the street, and the holy water fonts of churches. Its range of movement is limited; it can propagate the disease only when people live in close proximity. It is a relatively simple matter to spray urban dwellings with insecticide.

By 1909 a Mexican doctor, Eduardo Liceaga, was courageous enough to report "that the war on the mosquitoes is so efficacious that there are none left in Veracruz" (Liceaga 1910:63–5). This turned out to be somewhat premature, since yellow fever did recur, but in effect the pall had been lifted from the town. By 1924 yellow fever had been eradicated from Mexico through the destruction of its vector (Warren 1951:17).

Landmarks in the subsequent research on yellow fever were the confirmation of a virus as the active agent, the discovery of "jungle" yellow fever, and the development of a vaccine (Warren 1951:3–37). The "jungle" version of the disease merits additional comment. It is probably indigenous to the Americas and maintained independently of man among mammalian hosts, usually monkeys, in tropical forested areas. It is transmitted by a number of species of mosquitoes other than *Aedes aegypti*. Eradication is not practicable, as it is in the case of the vector of urban yellow fever; all the onus is on immunization. When unimmunized persons come in contact with these mosquitoes, say, during the clearing phase of shifting cultivation, the danger of the disease remains. If infected people then come into towns and are bitten by remaining or newly introduced *Aedes aegypti*, the urban form of the disease may flare up again.

NINETEENTH-CENTURY UNCERTAINTIES

Yellow fever was sometimes regarded as a type of malaria (e.g., Ruxton 1855:35). This word has its roots in the Mediterranean area (e.g., Italian: *mala aria*) and means bad air. It alludes to the evil that might be expected from the exhalation of swamps. There was a great deal of uncertainty about fevers generally, which is not surprising since it is still difficult to tell yellow fever from related conditions. The great range of severity was not recognized; only the more serious cases were considered as yellow fever. The typical course of the disease was not too clear

either. It could last a few days or a few weeks. The travellers did know
that one could contract it during the briefest of stays in town and not
know one had it until several days later. Even when already up above the
lowlands proper or on board ship headed home one could not be sure
for a while whether to chant *Te Deum* or *Miserere mei, Deus!* (Latrobe
1836:300).

There was considerable argument among medical men of the nineteenth century, especially in the United States, over the issue of conta-
gion (Bloomfield 1958:492–3). Humboldt observed that in Veracruz the
well were not afraid to approach the sick; yellow fever was not commonly considered to be contagious. His line of reasoning in support of
this came very close to a recognition of the necessity for a vector (e.g.,
Humboldt 1813, IV:403–5). In this and in other formulations regarding
the disease, Humboldt was considered by most of our travellers to have
gathered the prevailing wisdom and he was referred to without much
question.

The most interesting reading in the travellers' discussions of yellow
fever is usually their struggle with its cause, which was obviously a
challenge to logic and language. Humboldt had been eloquent in his
admission that much was still unknown about the origin of yellow fever
and related diseases (Humboldt 1813, IV:428). The miasmatic theory was
basic. A careful discussion of it consisted mainly of multiplying factors
and bringing them into some complementary action, of entering quali-
fications and performing circumlocutions. No one surpassed Humboldt
himself in this respect (1813, IV:388–403).

Yellow fever and related diseases were thought to be generated in situ.
They arose somehow out of the exhalations of rotting animal and vegetable matter. Its rank, suffocating smell signalled the danger. Any
accumulation of filth was thus indicted. Poor drinking water might be to
blame, although no one was quite sure why, except that it seemed filthy
and tasted bad. However, the finger pointed most insistently at the
swamps known to be out there among and just behind the dunes.
Intense heat, amplified by reflection from the slopes of sand, accelerated decomposition.

Ward raised an interesting doubt about this whole idea and per-
formed a neat evasion: "At Veracruz its cause has been sought in the
local peculiarities of the situation and there is little doubt that the
exhalations from the marshes which surround the town, must have a
tendency to increase the virulence of the disorder" (Ward 1828, II:244).
If that were so, he went on, how could one explain the appearance of
the disease wherever a number of Europeans had assembled for trade?
Some ten years later, Becher had similar doubts. Since the disease
occurred at Puente Nacional, surrounded by luxuriant vegetation and

passed by a substantial river, there had to be other sources for yellow fever than swamps (Becher 1834:36).

Generation in situ, generally accepted throughout our company of authors, put another line of thought into the background. Humboldt noted that it had once been assuring to put the blame for an epidemic on infection introduced from outside (Humboldt 1813, IV:384). There had traditionally been a great fear of infected ships, along this coast and elsewhere (Ackerknecht 1955:52). Veracruz and Havana had often accused each other of sending over some very unwelcome cargo. And there was a sound reason for fearing ships, although this was not known till much later. Infected *Aedes aegypti*, which might remain virulent for as long as sixty days, were often carried abroad in just this way. Jungle yellow fever is considered indigenous to the Americas, but *Aedes aegypti* are thought now to have been brought in by early slave ships, making urban yellow fever possible (Ward 1972:5).

All this about putrefaction and infected ships was brought together and given a particularly distasteful twist by Wilson, the American lawyer who set himself the task of exposing fallacies about Mexico more rigorously than anyone else had yet been able to do. Yellow fever was "an African disease, intensified and aggravated by the rottenness and filthy habits of the human cargoes that brought it to America" (Wilson 1856:24).

The accompanying oblique aerial photo shows the swamps that were considered by Humboldt and those that followed as chiefly responsible for the generation of the yellow fever that plagued the port (Figure 2). Ferry described them:

> The medanos (dunes) hinder the rainwater from flowing away. Small lakes are thus formed at the bottom of these sand-hills; and the parched-up ground is gradually converted into a fenny marsh, from which arise the most pernicious exhalations. A thick layer of mud fertilizes the sand, and all the noxious plants which abound in low, moist grounds are here produced in countless profusion. During the rainy season this rank vegetation spreads and grows round all the margin of the ponds. The mangroves shoot their branches down to the ground. They take root there, produce new trunks, and soon form impenetrable thickets—haunts of numberless reptiles of every kind. A thick crust of greenish scum carpets the surface of the water. The fermentation which sets in on the return of hot weather in these frightful marshes disperses deleterious miasmas abroad. (Ferry 1856:316)

Then as now, the dunes behind Veracruz were of various generations.

The aerial photo shows swamps at the junction of two generations, the earlier already fixed by vegetation of some complexity, the later still in movement. Dunes store surprising amounts of water. The swamps among them represent the intersections of the water table and the surface. As might be expected, the size of these swamps grows in the rainy season; they were thus most apparent when yellow fever was most threatening.

As early as 1885 a Mexican scientist showed the sort of interest in the dune biome that one might now expect from an ecologist (Ochoa 1885). This required courage and uncommon curiosity. Müller was similarly able to set aside the general fear of swamps when he passed through a larger wetland enclave to the west in 1856, as we will see. Ochoa was impressed by the storage of water in sand, the rich variety of plant species on the older dunes and the dynamics of the plant community in general. It was clearly evident, of course, that the dunes were being spontaneously fixed, but he suggested ways in which the process could be accelerated with useful plants. He cited the same miasmatic theory in explaining yellow fever and suggested it would be useful to prevent beach sand from drifting inland by erecting a kind of palisade. This would keep the animal matter that the sand continually carried with it from accumulating and putrefying in those malevolent hollows. Then, by covering the dunes with vegetation, the entire region would be "sanitized."

MORE CLEARLY UNDERSTOOD ENVIRONMENTAL CONCOMITANTS

It was well known that the disease prevailed during the summer months, but the onset might be March or it might be May. The *paz del Señor*, or the "peace of God," that is, the first *norte*, might come in September or November. Entire summers could pass without an epidemic and in bad years an epidemic might last the year round. The onset of the disease was a matter of rumour. Evidently the first question captains asked of the pilot was about the recent behaviour of yellow fever—the second about revolutions (Mayer 1844:3). The word that the *vómito* had begun might well set a man to packing his bags and leaving Veracruz precipitously for Jalapa (Ruxton 1855:32).

Air circulation was seen to affect the incidence of the disease throughout the coastal region. The first *norte* swept the contaminated air away. The effect of circulation was also noted at a larger scale. The walls of Veracruz were seen to constrict settlement and to inhibit the freshening of contaminated air among the houses on a daily basis. The houses themselves were too crowded. And, of course, it was assumed that somehow the miasmas were wafted into the town from the swamps where they originated.

All travellers knew that the disease had an altitudinal barrier. On the ascent toward Jalapa, this was signalled by the first oaks around 1,000 metres above sea level. On a descent the limit of the disease seems to have been sensed lower down, say, at Puente Nacional or even closer to Veracruz. The travellers were putting the threat off as long as possible. Ferry, the articulate vagabond, was finally brought up short by his Mexican manservant near Manantial. Cecilio would go no closer to Veracruz, and since Ferry did not have the funds to pay him until he could cash a "bill" (a letter of credit?) at one of the commercial houses in the city, they agreed to gamble over Ferry's excellent horse. Cecilio won and was well paid; had he lost, he would have paid a considerable price for his fear of yellow fever (Ferry 1856:280–7).

It has become apparent that both the periodicity and altitudinal limit of the disease are mainly a matter of temperature (Carter 1931:40–5; Warren 1951:3–37: Ward 1972). The time required for the incubation of the virus lengthens as the range of temperatures increases. With this, the effectiveness of the mosquito as a vector decreases. The *nortes* and an increase in altitude had their inhibiting effect, therefore, in that they affected the incubation of the virus. Temperature affects the epidemiology of the disease in conjunction with other factors. High humidity may lengthen the mosquito's life span. Rainfall provides countless small breeding pools within settlements and in the foliage of tropical vegetation. Winds also affect the direction in which mosquitoes on the wing are wafted. At all times there must be infected people from which to draw the virus and non-immune people in whom it can be propagated.

IDEAS ON SUSCEPTIBILITY

By the time our travellers were entering Veracruz it had long been known that yellow fever struck only the stranger. In parts of South America it was therefore called "patriotic fever" (Ackerknecht 1955:52). That hardly served along the Gulf Coast because Mexicans coming down from the Mesa Central were known to be as vulnerable to the disease as foreigners. Humboldt thought they were in even greater danger (1813, IV:408). The rapid change in temperature brought on during the descent from the higher country meant "all the pores are opened at once, and the general relaxation of the system necessarily renders [the uplanders] peculiarly susceptible of the disease" (Ward 1828, II:248). Foreigners arriving by boat, on the other hand, had time during their long voyage through the latitudinal belt to become accustomed to the heat.

Yellow fever was also thought to discriminate between the races. Some authors were convinced the Indian was immune (e.g., Mühlenpfordt 1844:1,351). Black people were generally considered less susceptible than whites or those of mixed blood. This point is not taken quite as far in the discussion of yellow fever along Mexico's Gulf Coast as it was in New Orleans, where it assumed a strikingly racist tone (Duffy 1966:143–5). The black man was different in his physical make-up and culturally inferior. Why not settle black people around the docks as a kind of buffer zone, to prevent the infection, which always seemed to come in by ship, from spreading?

The disease was particularly pernicious in that it favoured robust young adults over children or old people (de Fossey 1964:249; Ferry 1856:280). Humboldt seems to have laid the basis for this in his observations on the vulnerability of young soldiers posted to defence installations along the coast (Humboldt 1813, IV:408). It has had a recent echo in that the "jungle" version of yellow fever seems mainly to attack adult men, but then they are usually the woodcutters exposed to the mosquitoes that spread this form of the disease (Ward 1972:27). Incongruously enough, it was also thought in the nineteenth century that the delicate sex was less vulnerable than men. Nevertheless, the well-to-do insisted on hurrying their ladies out of town in the direction of Jalapa once summer came, or indeed, keeping them there all year round.

Some thought that the vulnerability was linked to class. Ward was convinced that a cultured gentleman could take precautions. He could avoid excesses, maintain a calm, controlled disposition, and rise above the whole sordid business: "The persons most likely to suffer would be servants, and persons of that class, who often will not be induced to prepare themselves for landing beforehand, and, when on shore, are either excessively apprehensive, or unnecessarily imprudent" (Ward 1828,II:249). *Arrieros* were of that class, and soldiers, crews of ships, even a group of French colonists arriving in 1826—all shown to have been decimated by yellow fever.

Another gentleman echoed Ward several decades later: "*Without care, or when arising from excesses which no rational man commits, no doubt the vómito is as dangerous and fatal as ever. For myself, I should not feel nervous about it, even in the sickly season*" (Robertson 1853, I:260–1)

One needed to remain calm, was the budding cultural anthropologist Tylor's advice. Once the disease had come for the summer, "the unseasoned foreigner had better lie on his back in a cool room, with a cigar in his mouth, and read novels, than go about hunting for useful information" (1861:21).

Acclimatization was known to make the most fundamental difference to those who needed to remain in the lowlands, although no one could clarify just how this was achieved. It was known that anyone who had been attacked and had survived was immune, but this was not linked with acclimatization, whereas it was precisely in such immunity that the key was eventually found to lie. The variations in vulnerability attributed to race, age, sex, physical condition, class or length of residence in the tropics, to the extent that they had any basis in fact, must have been mainly a matter of whole groups achieving immunity through widespread mild infection, often unnoticed attacks in childhood (Ward 1972:28).

Trying to bring together current ideas about cause and vulnerability, one author achieved an interesting summation. It was a consequential deduction from the literature on the disease up to that point: yellow fever arose independently out of the conjunction of constitution and place (Buschmann in Thümmel 1848:106). Such a deduction could not lead very far.

PREVENTION

What could be done to lessen and perhaps eliminate the danger of yellow fever? It was as clear to Humboldt, as it still is to any self-respecting real estate developer, that swamps needed to be drained. Such a remedy, however, was far beyond the resources of the people of Veracruz in the first half of the nineteenth century. It would be the second half of the twentieth century before they would get down to it with a will.

One could attack filth in the streets and this was done with considerable diligence. Humboldt advocated fighting like with like and fumigating churches, houses, and ships' cabins (Humboldt 1813, IV:428). Various more bizarre measures could be taken, such as direct attacks on the atmosphere. In the sixteenth century it had been the practice to drive two thousand head of cattle through Veracruz every morning in order to rid the town of the ill vapours of the earth (Chilton in Mayer 1961:18). Ackerknecht tells of how cannons were fired in Jacksonville, Florida, in the mid-nineteenth century in order to disperse the miasma (1955:57). Kenly, the Maryland volunteer in the Mexican war, saw the inhabitants of Veracruz keep fires burning in the streets day and night. He reflected on how nothing seemed to mitigate the threat: "The science of the ablest physicians is entirely at fault in dealing with it" (Kenly 1873:294).

The individual was counselled not to go out in the midday sun or to expose himself to the night air, two venerable precautions against a variety of things. One needed to eat right and keep regular, to avoid overexertion and agitation, and to insure that neither a porter who charged too much for carrying baggage nor a dirty hotel room, nor any other of the difficulties or fears associated with travel in Mexico, disturbed one's equanimity. This was very difficult, and for a moment someone got close to the truth: "The venomous insects that infest that hot region add likewise in no small degree, by never leaving the inhabitants to repose, and consequently causing feverish excitement and irritation, to promote disease" (Gilliam 1847:24).

It was clearly very important to avoid tension, but one should also avoid the other extreme, excessive relaxation of the nervous and vascular systems (Duffy 1966:149; Sartorius 1858/1961:3–4). In this lay an old caution about the tropics. They are seductive; one must maintain reserved habits, and keep the moral fibres taut.

It is easily imaginable that the various contradictions in the wisdom of the day about this whole horrible risk will themselves have brought profound unease to anyone who thought at all about it. Fortunately, there was a last resort: "No care, no precaution, no previous course of medicine—no certain antidote can be prescribed. In daring it from necessity, you must rest satisfied with following the advice given, and taking those measures, which, however vain in many cases, experience has sanctioned, and throw yourself upon the mercy of God for the rest. . . . And this we had done to the best of our ability" (Latrobe 1836: 300).

The same author tells the story of the death of a young man, which becomes a parable (304–9). Gay, careless, French, without religious belief as it later turned out, this young man spent several days in Veracruz socializing with acquaintances. Like other passengers coming down from the escarpment to embark in Veracruz for Europe, he should have stayed in Jalapa to the last possible moment and then proceeded expeditiously to the ship. He came on board on the eve of departure, "dilating upon the social hours he had passed in consequence of his better management. Poor fellow!—little did he imagine, that that heedless contempt of danger would cost him his life; that at that very moment, the seeds were sown of the fatal disease—and that in the eyes of more than one experienced observer on board, he was already a doomed man!" (305).

Of course, the poor fellow did fall prey to a classic, severe case of yellow fever. Everyone did the best they could; but the sufferer entered delirium and the ship a storm. He succumbed on the eight day out and was buried at sea with Church of England ritual.

TREATMENT

Advice was traded back and forth on this subject, and various degrees of effectiveness claimed, but it was usually admitted that the risk of death remained high, certainly in a severe case. Occasionally someone reported a sure cure, but nothing had a significant impact: "The general terror of being sent to hospital is so great, that many are deterred from applying for relief until their cases are beyond the range of remedies" (Thompson 1846:5).

Two main trends are apparent in the treatment. The one, used widely during epidemics in Veracruz, was heroic medicine. It involved, among other things, the administration of huge doses of calomel, a drug containing mercury. Only when patients showed symptoms of mercury poisoning did they seem to be getting real benefits from the treatment. Little by little, a very different strategy took hold throughout the yellow fever areas around the Gulf. Newly trained doctors returning from France were "convinced of the value of leaving the patient alone and permitting nature to work the cure" (Duffy 1966:152). That seems to be the modern approach: doctors are only encouraged to apply various palliatives (Krupp and Chatton 1977:796). There is no cure: the emphasis is on prevention through immunization.

THE EFFECTS OF THE FEAR

In practical terms, the fear of yellow fever complicated the defence of the port and the coast. This had been so throughout the colonial period but especially when England threatened the eastern shores of New Spain late in the eighteenth century (Archer 1971). The population of the coastal region was not large enough to provide a sufficient defensive force. Troops brought from the interior ran a frightening risk and resisted transfer to Veracruz by all available means. The result was a great deal of official agonizing, considerable paperwork, and chronically under-manned garrisons. Similar problems and a high mortality rate among Mexican soldiers were noted by nineteenth-century authors too.

It was observed how fear of the disease could create unrest among the population. The Veracruzanos evidently had a very simple saying: "¡septiembre, setiemble!" that is, "In September one trembles." The reference was to the most feared latter part of the wet season, by which time the miasmas had built up their greatest virulence. Real terror, such as was experienced in the Mississippi Delta in 1853, emptied cities and villages, committing greater ravages, it was judged, than the pestilence itself (Wilson 1856:24).

It was not difficult to see how the threat of yellow fever impeded commerce. Veracruzan trade was subject to the periodicity of the disease. Ship captains preferred to arrive during the winter in order not to lose crew members or to have to submit to quarantine on return to Europe. The effect on supply and prices in the trading houses of the port can be imagined. *Arrieros* would not approach the town when an epidemic was on and so the movements of goods by land was slowed. This in turn would have its repercussions in the capital and in the mines. Humboldt had laid all this out pretty clearly, and had also recounted the argument among government officials during the time of his visit about the advisability of simply destroying Veracruz altogether and settling everyone elsewhere. Nothing came of this, nor could it; there were sixteen thousand people, some very prominent families, and a great deal of capital investment involved (Humboldt 1813, Band IV,376–7). The numbers of traders who, year in and year out, took on the personal risk of death as well as all the practical difficulties that were associated with commerce in Veracruz are an indication of just how good the prospects must have been.

For the travellers the menace that lay in wait was unsettling. Madame Calderón de la Barca, quoting Milton, had described the *norte* as the flaming sword that guarded the entrance to paradise (Calderón de la Barca, Fisher ed. 1966:50). Justo Sierra Mendez, a Mexican writer of note and a traveller, who crossed the lowlands by rail in 1869, made of yellow fever a more sinister guardian. It was like the classical dragon that guarded the golden apples in the garden of the Hesperides (Tavera Alfaro 1964:370). Both images are interesting in that they postulate a fabulous interior. The acerbic lady was probably being ironic, but most of our travellers did expect to find the fabulous—in nature and perhaps by way of economic opportunities as well. The dragon fits well among the dunes beyond the walls of Veracruz. Its presence there was hardly conducive to deliberate preparation for the journey inland or a good night's sleep, even if one did not have to stay in a hotel. The commentary on the town is hurried, the forms and tones surreal.

POSTSCRIPT

It is to be expected that the foreigners' fear of yellow fever should mask the year-round dying among the residents of Veracruz. The one breakdown of a year's death toll in Veracruz that is available is intriguing, and as limited as some of the other isolated tables of statistics ferreted out by especially diligent or lucky travellers (Mayer 1844:7). Six major categories of fatal diseases may be abstracted from it, and yellow fever is only in third place among them (see Table 5).

Several consideratons must be borne in mind. On the one hand, *vómito* was probably more important than the entry indicates, because it could be confused with other fevers and diseases of children. On the

TABLE 5: Deaths in the four hospitals of Veracruz, 1841

Phthisis* and diarrhoea	212
Diseases of children	162
Vómito	155
Smallpox	142
Fevers	142
Convulsions	50
23 other categories with small numbers in each	154
	1,017

*Pulmonary tuberculosis

other hand, the *vómito* figure no doubt includes foreigners and upland-ers so that the significance of the disease for the inhabitants of the town diminishes again. All in all, the deaths attributable to poor hygiene, poor housing, and poverty overshadow the toll of the evil exhalations. From behind the preoccupations of the travellers emerges again the more routinely hard-pressed town.

CHAPTER FIVE

"Hay que armarse para viajar"

Departure from Veracruz would be a relief, promising freedom from the danger of yellow fever within perhaps a day's travel, but it would also bring on the rigours of the road: heat, thirst, primitive lodgings, troublesome dealings with servants and innkeepers, as well as a further major hazard, highway robbery. The approach to the harbour had been menaced by the *norte*—a matter of luck. The stay in town had been imperiled by an utterly sinister disease. Robbery was at least a dignifying danger; against it one had a sporting chance, one could arm oneself. One traveller put it in stronger terms, "Hay que armarse para viajar" (de Fossey in Glantz 1964:255). To travel in these parts, he judged, one had little choice but to arm oneself.

Both the rigours of the road and its insecurity became part of the indictment being formulated against not only the tropical lowlands but the entire country. These were all aspects of its disreputability. However, the possibility of an encounter with highwaymen in some desolate place also stimulated romantic sensibilites. And it called forth a certain truculence, which may be seen as the focusing of a common attitude toward the country as a whole and an expression of the policy toward Mexico of the countries from which the travellers came.

MEANS OF CONVEYANCE

Just for the record, as it were, Mühlenpfordt informed his readers that only the Indians travelled on foot. Only they could endure walking through the heat of the lowlands and the thin air of the mountain passes. Most other travellers rode a horse or a mule (Mühlenpfordt 1844:1,341). Arrangements varied according to means, but one usually did not go alone. A mounted servant followed close behind and perhaps a spare mount, as well as a mule with baggage. Foreigners commonly purchased their mounts for the duration and then resold them. The servants were usually engaged in Mexico and there are several accounts of very beneficial attachments thus formed. Those brought from abroad were not of much use: they usually did not speak Spanish and in any case might well fall to pieces under the challenge of the new circumstances (Ward 1828:II,178). This way of travelling was general throughout the country. Even the wealthiest male travellers, foreign or native, commonly went this way, at least wherever stagecoach service was not available (Ruxton 1855:89).

The mounted traveller could move independently, and rapidly if he wished. There was greater exposure to the elements, but a hat and a headcloth protected him against the sun and a serape against the rain. He could stay on the main road or go off on lesser trails. This made him an elusive target for the robbers, but more importantly, it was the best way of seeing the country and, indeed, the most sensitive accounts of land and life in the tropical lowlands come from those who travelled in this way. Heller, for example, was unaccustomed to riding, poorly mounted, and rattling with armament, all of which reminded him of Don Quixote (Heller 1853:39). But he was enthralled with tropical vegetation and described it well. Staying on the main road was for the uncurious, Sartorius maintained (1858/1961:4). Others were likely to be frustrated. Coaches travelling at night, for example, sometimes came on a fiesta in some wayside village. Suddenly an illuminated porch with dancers! The stupefied passengers would stir and just as they began to appreciate the scene, the driver's whip would make of it an illusion (e.g., Vigneaux in Glantz 1964:498).

To leave Veracruz first class one hired a *litera*, a conveyance that would have created a sensation on Broadway, as Mayer maintained (1844:7). It was, in effect, a canopied couch (Figure 7). Its two carrying poles passed through traps on the backs of mules, one fore and one aft. Two drivers, and perhaps a pair of mules for relief, were part of the bargain. The *litera* was relatively safe because the single passenger, or at most two, could take few possessions with them. Luggage and valuables were usually shipped by well-guarded mule train. There was therefore little to be gained by holding up such an affair. However, it needed at least twice as much time to cover a given distance as a stagecoach and was much more expensive per passenger. On straight ground it was marvellously comfortable, in fact one traveller disdained it as a boring, feminine conveyance (Koppe ed. 1955:82). On steeper, rocky stretches it pitched and rolled like a ship at sea.

Light, wheeled vehicles of various kinds were used by several prestigious parties. Poinsett's group rented *volantes*, small two-wheeled affairs

7 Various means of travelling in Mexico, in Wilson (1855:284)

with leather suspension, drawn by three mules (Gardiner 1959:21). Koppe had information that they were prone to upset—more than once each day (Koppe 1955:81). Ward's party brought three of their own four-wheeled English carriages, which they later wished they had left behind, since the vehicles were intended for urban use and were not rugged enough for the Mexican roads (Ward 1828, II:179). Occasionally, a traveller might charter a coach, just to have some room for his baggage and the right to start and stop at his convenience, or even to go off the main road (Koppe 1837, II:306).

Private parties, and often stagecoaches too, were provided with mounted escorts under a variety of arrangements. Almost without exception the travellers did not find these guards to be very imposing specimens: neither their horses nor their arms or demeanour inspired confidence.

Regular stagecoach service was established between Jalapa and Mexico City in the late 1820s by an American company which was later bought out by a Mexican. It operated with imported American vehicles, drivers, and methods. Indeed, it became a symbol, a marvel to Mexicans and Europeans, and a source of pride for Americans themselves. Here was a system that worked well and, like every other innovative public service in Mexico, Richthofen maintained, it had been established by foreigners (Richthofen 1854:376). However, there was something particularly headlong and forceful about this service.

American coaches were preferred from the time the service started until its eclipse by the railway. They are described in many accounts, and occasionally drawn. They were not as heavy, slow, and bumpy as their Mexican counterparts, yet they were rugged in their construction, especially the woodwork. Their suspension system was of leather; they had larger wheels on the back axle than on the front in the interests of stability, good braking, and manoeuvreability. It is likely that the carriages described included examples of the famous Concord coach (Wilson 1855:41; Time-Life 1974:136-7).

More impressive than the American coaches, however, were the American drivers. They seemed fearless, were able to manage the unruliest teams and achieve amazing speeds. Coming down the escarpment into Jalapa with them must have been a memorable experience. The best seat on such a ride, at least in good weather, was up beside the driver—a sensation no longer available on any modern conveyance.

During the early years of this service it went only between Mexico City and Jalapa. The traveller arriving in Veracruz had to find some other way of getting to Jalapa; often this was by *litera*. By the late 1830s there was service from Veracruz three times a week. By the mid-1850s, the frequency had doubled. The journey from the coast to the capital would normally take four days and one could expect to travel two out of the three nights.

Altamirano, the Mexican traveller and writer of note, recalled how he saw the coach with youthful eyes as it rolled along the horizon, and remembered what it had meant for his country at the time (Tavera Alfaro 1964:298). It was elegance and modernity; it left behind the earlier heavy coaches which had been slow as litigation over a will. It was shown in advertisements as riding on clouds with the wings of Pegasus. Provincials hoped some day to be able to afford a ride on it, and people of means preferred it to any other means of travel.

The foreigner leaving his inn at a horribly early hour saw in the courtyard an ungainly looking vehicle, even if it was a Concord. Comparisons were being made with lighter European carriages. Moreover, it was usually battered, and the mules that were to pull it were not very prepossessing animals. Respect for this conveyance, and an understanding of just how rigorous travel in it was, would come later. To take one's seat in it meant turning the clock back fifty years; to take the beating that was about to start one had to be in good physical condition (Wappäus 1863:93).

THE ROUTE

The road under construction when Humboldt travelled to Veracruz was to serve as the principal connection between the capital and the port for more than half a century (Figure 1). It represented one phase in a long rivalry between the Orizaba and Jalapa routes—a temporary resolution in favour of the latter. The subsequent reversal with the completion of the railway between Veracruz and Mexico via Orizaba has been outlined in Chapter 1.

Both routes had their antecedents in pre-Columbian footpaths. Cortés took the trail through Jalapa in his ascent to the plateau, but he passed between the peak of Orizaba and that of the Cofre de Perote, not north of the Cofre along the present route (Rees 1976:27). Sandoval took the Orizaba route in his expedition to suppress an Indian uprising around that town and to found Medellín (Mayer 1961:65). It was longer and failed to achieve the significance of the Jalapa trail for long-distance transport during the colonial period.

The first wagon road from the plateau to the port was planned for the Orizaba route. Around 1590 Juan Bautista Antonelli proposed a connection between the capital and the Venta de Buitrón on the present site of Veracruz, to which the port was to be moved from La Antigua before the century was out (Figure 1). The idea was rejected, mainly because it seemed to Viceroy Velasco that it would be quite impossible to marshal the labour to build such a road (Melgarejo 1960:97; Rees 1976:34). Nevertheless, a fine map in six colours was deposited in the archives (Mayer 1961:62-3). The modern expressway from the capital to Cordoba approximates Antonelli's route.

New plans were laid for wagon roads along both routes in the late eighteenth century (Melgarejo 1960:98-9; Humboldt 1813, IV: 286-7). The merchants of Mexico and the Orizabeños were in favour of the one. The merchants of Veracruz, many of whom had residences and trading connections in Jalapa, insisted on the other. Viceroy Iturrigaray resolved the issue in 1803 in favour of the enterprising consulado of Veracruz by decreeing them a suitable financing formula and putting a capable engineer in charge, Captain Diego Garcia Conde (Trens 1955:45). Roadbuilding continued until the beginning of widespread unrest in 1811, but the project could not be completed. The village of Santa Fé was left as the eastern terminus, rather than Veracruz (Figure 1).

In the meantime, work was progressing on the other route as well; the stretch between Orizaba and the foot of the last steep ascent to the plateau, just above a town called Acultzingo, was completed before the struggle for independence began. Short segments of the old stone pavement were still visible until recently alongside the slightly smoothed but still quite unnerving curves of that ascent. In the 1830s all this was certainly in as bad or worse condition than the road through Jalapa, and much less used (Mühlenpfordt 1844:II, 64).

The route between Jalapa and the coast changed several times in response to the southward shifts of the port. The pre-Columbian route used by Cortés left Villa Rica, the earliest and northernmost site of Veracruz, passed in a southward loop through Quiahuiztlan, Zempoala, Rinconada, and then on via Jalapa and Perote, as in Figure 1. It was made more direct in two stages. In 1524 the traffic was redirected from Rinconada eastward to La Antigua, the new port, and from there southward along the beach to Vergara and the Venta de Buitrón on the site of the present Veracruz. Most ships anchored opposite this point even though the official port was to the north, in order to exploit the shelter of the island on which the fortress of San Juan de Ulúa was eventually built. It was not much of a shelter, but the best along the coast. For some time La Antigua was in competition with Medellín, the village to the south of Veracruz, on the Jamapa River, that was to become a resort for Veracruzanos. Some trade was enticed in its direction. By 1600 the issue had been settled in favour of the present site of the port, and the route could be shortened from Rinconada through Paso de Ovejas, to Veracruz, more or less the route one still follows today (Melgarejo 1960:96; Rees 1976:29).

Condition of the Road

The first segment of the road out of Veracruz was the smoothest part of the journey. From the northwestern gate it went out along the beach. Passengers awaiting disembarkation could see the line of traffic along the horizon to their right, and wish they were already on their way as well. A view of this stretch of the beach, with the town and the fort in the background, was painted by Thomas Eggers in 1810 . The light suggests morning, the sea is out, and the sand is shining. There are no coaches in sight; they were probably quite rare in this area at the time. A rider is galloping toward the town, perhaps carrying the mail. A line of pack-mules is coming in the opposite direction. Fishermen are busy around beached sailboats. Just metres from the waves a *litera* is moving by, bound for the interior. The curtains are raised to allow the sole occupant full benefit of the breeze.

Just before the beach curved northward the road veered off to the west. It was one of the most difficult parts of the road for the draft animals. Wheeled vehicles often sank to the axles. Various writers refer to moving dunes that obliterated routes from one year to the next. This was the desolate part, with nothing but the skeletons of mules and horses to show it had been traversed (Bullock 1825:31). Koppe came on something more: in 1826 some English investors in Mexican mining had sent out a rather large steam engine which had got no further than a short distance into the dunes, where it was gradually being covered up (Koppe 1837, II:313). One writer refers to the "callejones de Santa Fé," sunken, sandy roads lined with scraggly vegetation (Blanchard 1839:91). Once out of the dunes and past the village of Santa Fé, the journey began in earnest on the road built shortly before Independence.

This road was under construction when Humboldt came down the escarpment in the spring of 1804. He predicted that it would become a marvel of engineering, comparable to the best mountain roads in Europe (Humboldt 1813, Band IV,288). And indeed, by 1811 all the necessary bridges had been built and a roadbed of rounded stone, some of it surfaced with cement, was in place from Santa Fé to Perote (Melgarejo 1960:99; Rees 1976:80). However, as the century went by, our authors found it more and more difficult to say just how badly this once so imposing facility had deteriorated, just how terrible it was to travel over it in a stagecoach. One traveller who endured the journey in 1825 observed that if a shake-up was good for a bilious constitution, then this was the best road to health (Hardy in Glantz 1964:121). In any other country, Mayer felt, it would be called the bed of a mountain stream (Mayer 1844:13). Until breached or eroded the cemented surface allowed quite comfortable travel. Thereafter the stones administered a jolting beyond belief, which can still be approximated in an automobile on those bits of the old roadbed accessible above Jalapa near San Miguel. During the post-1811 engagements between royalists and insurgents the roadbed was cut in many places. Wheeled traffic was thus virtually precluded and communication interrupted, which could be an advantage for either side. The loosened stones were often piled up for breastworks. The subsequent traffic encountered rutted interstices. Washouts were a continuous problem, tearing at the cement and exposing the stones. The pounding of hoofs and wheels left sharp holes.

In their own interest, owners of the coachlines from time to time had some of the worst spots repaired. Little was done in the way of maintenance by any governmental agency. Although tolls were levied by various authorities, and some of the money realized was intended for maintenance, little was in fact ever achieved, which is not surprising in view of the other priorities and the general administrational chaos that prevailed throughout our period.

Prescott had heard about this road, and although he was tempted to visit Don Calderón de la Barca and his personable wife in Mexico City, he demurred, fearing among other things that he might break his neck (Woolcott 1925:150). This was no exaggeration. Vigneaux told of how his coach oscillated irregularly enough to induce sea sickness, and how at especially deep holes, it threw one person against the ceiling and another into the laps of his travelling companions. Meanwhile it was infernally hot and everyone was thirsty. There was nothing to do but to sing boisterously (Vigneaux in Glantz 1964:497). One could take a different approach and regard it as the thrill ride of one's life, especially with an American at the reins. Koppe did that, as we have noted, making sure he got up on the driver's seat. Müller had a similar ride. He found he had to rig a kind of seat belt or he would have been catapulted. Beside him sat the driver's helper, whose main job it was to pitch stones from out of a bucket down on to the backs of lagging mules. In his own mind Müller rehearsed what he had heard regarding accidents on journeys of this sort (Müller 1864:212).

The fact that an engineering marvel should have become a ruin so soon and remain one for so long was a vast object lesson. It confirmed what the travellers had noticed in their first look at Mexicans: they lacked discipline and real competence. For a moment one could grudgingly credit the Spaniards for an achievement and let the Black Legend descend fully on those who had inherited New Spain.

Implications for Freighting

Freight cartage, the prime objective of the initiatives of the Consulado de Veracruz, was even more seriously affected by the poor post-Independence condition of the Jalapa road than passenger traffic (Richthofen 1854:380–1; Pferdekamp 1958:69–70). The common vehicle was the four-wheel cart, drawn by eleven mules, twenty-two on mountainous roads. Two-wheeled carts are sometimes referred to as well. In 1841 Mayer reported what must have been an amazing sight: fifty wagons laden with heavy machinery for factories near Mexico passing them somewhere between Plan del Río and Jalapa (Mayer 1844:13). Mining companies working in the interior did manage to get huge steam engines, pumps, boilers, and ore mills through the dunes and up the escarpment (Pferdekamp 1958:103). The cost of industrialization by these means, of course, was extremely high.

The normal dry season trip of a train of carts between Mexico City and Veracruz, say the trip of the guarded train that brought bullion to Veracruz, required from twenty to thirty days. De Fossey, who had elected to travel on horseback but as a tag-along to a train of carts because such trains were seldom stopped by bandits, noted that the strain on harnesses and wagons was always severe and that breakdowns were frequent. At the best of times the pace of the train was terribly slow. There was always plenty of time to ride ahead to buy provisions or go on excursions to the left or right (Glantz 1964:253–4).

Freight wagons were actually prohibited on this road during the rainy season because of the way their wheels damaged the unpaved sections, but the ban was not rigorously enforced. In the rainy season the trip could take three months, and any machinery transported then would arrived rusted.

In 1824 some English investors were prepared to rebuild this road and approached the federal government for permission to do so. They were turned down at first. When the government later changed its mind and would eagerly have accepted their proposal, they had lost interest (Mühlenpfordt 1844,II:64). By 1828 the initial optimism of English investors toward Mexico generally had been dissipated (Ward 1828,I:iii-vii).

Throughout our period most of the freight still moved by the traditional colonial muletrain or *atajo*, conducted by the *arrieros*. The poor condition of the road slowed this traffic also. The ruts dug by the wheels of the freight carts and coaches were obstacles to the animals. In fact, muletrains were better off on unpaved trails where carts never went and where heavy rains had less effect. But somehow the mules always got through. This was translated, of course, into high freight rates. Richthofen noted in 1854 that these costs were greater on many imports than the original purchase price in Europe. And this was not something that affected luxuries alone; most manufactured articles needed for everyday life had to be imported (Richthofen 1854:382).

Muleteers were the true fraternity of the road. All travellers met them coming or going and many had reason to look carefully at how they managed the transport of freight. Passengers in the stagecoaches and *literas* usually shipped their excess baggage and valuables by muletrain. Those travellers who represented companies, or thought they might have something useful to say to prospective colonists or businessmen on their return, observed how the freight was normally handled and how it could be optimally packed.

Each mule could carry about three hundred pounds and with such a load traverse five or six leagues per day. Less weight meant a more rapid journey. A load was best divided into two bundles, hence it was a good idea to pack goods in bundles of no more than one hundred fifty pounds. Anything too heavy or awkward would annoy the *arrieros* and slow the mules. Personal effects brought to the port in chests might well have to be entirely repacked (Ward 1828,II:263).

Muletrains travelled at their own rhythm, stopping for the night when the day's objective was reached, at some place where there was water and pasture. The *arrieros* usually camped out in the open, improvising sleeping shelters with packsaddles and mats in case of rain. Sometimes they settled in around a *venta*. This was the rudimentary shelter that served mounted travellers, or even those going by *litera*, for an overnight stop when neither Jalapa nor Veracruz, with their hotels (called *mesones*), was in reach. Each halt brought a standard ritual for the *arrieros*. First the animals were unburdened, then they were watered and fed some corn, after which they were freed to graze wherever this was feasible. In the meantime, someone would build a fire and begin the men's supper, which usually consisted of some combination of tortillas, chile, beans, rice, salt, and dried meat, all of which was carried along. After the meal everyone rolled up in their serapes and bedded down around the fire. Before dawn the camp was astir. The most demanding chore was reloading the mules. Mühlenpfordt watched such a scene and was especially impressed by the way two *arrieros* would pair up to tie down a saddle (Mühlenpfordt 1844:I, 341–2). The girth was tightened so hard that it dug an inch into the stomach of the mule. This was obviously a key prerequisite for a good day's travel. The tortured mules coughed and the *arrieros* swore. They had a special command of the language and any traveller who tried to describe them had to refer to this talent. They were rustic, expert, cruel, often characteristically and colourfully clad. Their mules might well be decorated too, with bells and inscriptions on their harnesses, very much as truck drivers today sometimes ornament their vehicles. It was well known that the *arrieros* were armed, and hence their *atajos* were seldom robbed. Once they got into town, of course, they became a rowdy, brawling bunch. One thing intimidated them, as we have seen: word of an epidemic of yellow fever in Veracruz. There is also a hint that they were not free agents. Their freight contacts were arranged for them by brokers (Richthofen 1854:381); they were probably tied to their job by indebtedness.

A traveller struggling with the problems and costs of arranging the shipment of this baggage from Veracruz into the interior expressed what must have been a common thought. The year was 1866; a rail line to the capital had been begun. Once it was finished surely all these inconveniences would be a thing of the past (Domenech 1922:29).

PRINCIPAL HAZARD

Travelling into the interior was not only inconvenient, it was also dangerous. Everyone undertaking it had been amply forewarned by tales of robbery on Mexican roads circulating in Europe and North America, among the passengers during the long crossings, as well as in the hotels of Veracruz. They were to be recounted more urgently than ever in the stagecoaches themselves, where they constituted a kind of "conversación sacramental" (Tavera Alfaro 1964:16). For the traveller, this was the most immediate aspect of the post-Independence judicial condition of Mexico, one further aspect of the deterioration already evident in the physical conditions of the road.

Koppe and Richthofen have supplied us with some useful assessments of Mexican justice in the early nineteenth century. Both authors were diplomats and both had some legal background. That of Koppe was probably the more practical since he was also the representative of a German company. Both of them quoted from Mexican justice ministers' reports in order to make some of their strongest points. Their interest was stimulated to a great extent by what they knew of highway robbery. Both used a word that will have been as sobering to them as to those of their countrymen who contemplated immigration or investment: *Willkür*, arbitrary behaviour.

Mexican law at the time was a mixture of holdovers from the colonial past and unintegrated subsequent legislation. Contradictions abounded, much was unwise on various grounds, out of date or just simply unworkable. Judges were hopelessly overloaded, distracted by conflicting interests, and often bribable. Lawyers could easily exploit the situation in order to manipulate judgments. The resulting court proceedings, of which Koppe seems to have attended a good number, were sometimes infuriating and sometimes just humorous. He thanked God that nothing like that had been heard in German courts for many years (Koppe 1837:I,246–57).

Evidently there were some barbarous old criminal laws still on the Mexican books. It was judged that up to the end of the colonial period these were enforced sufficiently to keep banditry at a minimum, at least along this road, and that after Independence enforcement deteriorated. The apparatus of justice hardly penetrated outlying areas at all. For this reason, above all, Koppe discouraged German agricultural immigration. Businessmen were not as vulnerable as pioneers: they would know how to protect their interests. They had consuls and ambassadors to help them get redress. Prospects for them in Mexico were in fact presented attractively enough by Koppe and other writers to ensure a substantial and respected German presence in Mexican commerce by mid-century. Other visitors had best beware; by venturing on to Mexican roads they were moving beyond the reach of the law, with little more to protect them than their wit and their arms.

Robbery had its own geography along the Jalapa Road. It was maintained that bandits had their favourite haunts at river crossings. Wilson's drawing of the stereotypical coach robbery is set at the crossing of the Río de la Antigua, with the famous bridge in the background (Figure 8). Other crossings, such as those at Paso de Ovejas and Plan del Río, presented similar opportunities for ambush. In reality, the low, spiny forest vegetation that predominated all along the road westward of the wetlands almost to Jalapa provided no end of such opportunities, as the American soldiers were to find out when they were harassed here by guerrillas in 1847.

Moreover, there are deep barrancas roughly parallel to the road on both sides for most of its descent from Jalapa to the wetlands. The barranca of Central Veracruz invites more careful historical geographical study than it has yet received. It has its own altitudinal zonation; it offered water, a variety of agricultural possibilities, and cliffs of conglomerate rock for isolation. All in all, it was an excellent refuge not only for nineteenth-century bandits and guerrillas, but for escaped slaves before that, and for revolutionaries in the early twentieth century.

The rocky, pine-clad slopes just above Jalapa were thought to be especially dangerous. Any steep grade, where the coaches had to slow down, was an invitation. Modern truckers still need to watch for hijackers at exactly these places. The most famous robbers' nest of all was just beyond the summit and before Perote, where rugged topography fringed the widening high plain. Many more such spots awaited the traveller between Perote and Mexico City. The landscape itself, one author reasoned, tempted its people to banditry (Bullock 1866:32). If one now considered that like as not these choice spots would be passed in the dark, the potential for some excitement was considerable.

Everyone had a fairly good idea how a typical robbery should proceed. Two Spanish words, frequently misspelled, were standard in this regard: *boca abajo*, that is, mouth down. Bandits would make the pas-

8 The robbery, in Wilson (1855:40)

sengers get out and lie with their faces to the ground. It was arguable whether or not they might show special consideration for priests and women. All would be robbed of money and valuables regardless. The possibilities for sardonic exchanges between robber and robbed, of course, would be endless, to be embroidered and enjoyed in the retelling. One such exchange is said to have taken place between a rich lady and a robber who demanded her ring. It would not come off. The robber threatened that it was either the ring or the finger, and the ring came off.

A good deal more may be gathered about the nature of this activity from scattered observations, but especially from some plausible reflections of Ruxton, and a rather careful analysis by Richthofen (Ruxton 1855:39-42; Richthofen 1854:132-5). The bandits clearly had the road subdivided into territories and worked them systematically. The coach drivers were known, and the behaviour of driver and passengers was remembered from one run to the next. For this reason the driver of the coach could not help much to prevent a robbery. He knew that if he did not stop promptly when hailed and become neutral during the proceeding, he or one of his colleagues would be shot peremptorily on the next run. Informers out of the communities from which the bandits came would be treated in the same fashion.

The guards were also often suspected of being in league with the bandits, which was logical. It might well happen that before an ambush they would fall some distance behind and regrettably arrive at the scene just after the deed was done. The guards also seem to have acted as scouts. It must have been disconcerting in the extreme to see one of them carefully inspect the passengers and baggage at a rest stop, counting the arms and assessing the goods, and then galloping off into the night (Ruxton 1855:41-2; Thompson 1846:20).

The guards did have their uses. They earned money by "guarding" and could be expected to have some arrangement with the bandits so that these would desist sufficiently to make such protection minimally believable. In any case, no one would want to carry banditry to the extent where it would seriously diminish traffic. And it was better for the passengers to pay these gentlemen a reasonable sum, it seemed, than to argue with them and risk an unpredictable encounter with their cohorts farther down the road (Robertson 1853,I:308-9). The whole activity could be seen as a system of tolls, an exploitation of a renewable resource.

The phenomenon was given several more sinister aspects. Convenient arrangements of all sorts between bandits and politicians or army officers can be imagined along the Jalapa Road or elsewhere. Beltrami, for example, was the victim of a group of highwaymen who, he was convinced, were in league with the governor of the state of Tlaxcala, from whom he had had friendly advice as to which route he should take to avoid bandits (Glantz 1964:237–8). Valois was given information about a protection racket operated by well-organized robber bands for certain merchants in the port. The band leaders were, it seems, eagerly invited to these merchants' tables (Valois 1861:72). This was a further reason why the cart or muletrains were so seldom robbed, and that single travellers were prepared to join such trains and take the slow pace along with the safety.

Many foreign visitors thought about the reasons for the whole troublesome issue of banditry in Mexico, some more incisively than others. In at least five accounts robbery was taken as an activity to which Mexicans were naturally inclined: "With the general population of the country lazy, ignorant, and, of course, vicious and dishonest, there is no lack of recruits for the road. . . . Perhaps the most powerful incentive to robbery is to be found in the insatiable and, as it would seem with the lower classes of Mexicans, constitutional passion for gaming, and the entire absence of all restraint in its indulgence" (Thompson 1846:23).

There was an actual instance reported where temporarily embarrassed gentlemen took to the road (Thompson 1846:24). After the joint British, Spanish, and French expeditionary force had landed in Mexico in 1861, there were also reports of how French troops out of the Second African Regiment and the Foreign Legion had joined the ranks of the brigands (Bullock 1866:8). Some attributed banditry to the breakdown in law enforcement after Independence. During the time of the Spaniards, a caught bandit never escaped the rope; now that deterrent was missing (de Fossey in Glantz 1964:256). However, the underlying opinion, implicitly or explicitly, seemed to be that robbery by Mexicans was "a sin peculiar to themselves, which should cover them with shame and the contempt of the world" (Gilliam 1847:35).

The theme of the Mexican as a bandit circles through the discussions of robbery in this literature; it was obviously an idea well-rooted abroad, particularly in North America. It devolved especially on the soldier. The stagecoach guards looked like bandits and were often in league with them. The Mexican soldiers that fought the American soldiers in 1847 did not fight openly as honest men should, but harrassed the Americans from behind impenetrable vegetation. This was the "guerrilla system," tantamount to brigandage (e.g., Kenly 1873:301–2, 311–12). Years later, "revolutionary" became virtually synonymous with bandit. An American army man and foreign correspondent, for example, writes from Mexico in 1912: "With the collapse of Diaz's rule last year came the immediate return of highway robbery and brigandage all over the country. For a while the bandits practised their robberies under the pretense of being honorable revolutionists; but after the triumph of the Madero revolution, when no such excuse was left, they threw off the mask and became brigands pure and simple [including Emiliano Zapata] . . . the greatest bandit leader of the south" (Emerson 1912:233).

Richthofen analysed banditry in what one might consider structural terms (Richthofen 1854:132–5): Mexicans were not naturally inclined in this direction; in fact, climatic conditions tended rather to leave them soft and apathetic. He went on then, a little more plausibly, to search for the roots of the phenomenon in the colonial period. This line of enquiry, of course, has since been taken with respect to many socioeconomic conditions in Spanish America (e.g., Stein and Stein 1970). Many industries, crafts, and agricultural pursuits had been proscribed in order to enhance metropolitan interests. Much of the economic activity that was lawful was encumbered with regulations. Interregional trade was burdened with very troublesome duties. On every hand the venturesome were hindered. To avoid hindrance, in fact to survive, one had to break laws. To circumvent became honourable and to deceive necessary; forbidden economic activities were freely practised. From the readiness to violate governmental regulations to a forcible move against others' private property seemed to Richthofen only a step. Such criminality, he had concluded, could still be held in check by strong enforcement during the colonial period, but he maintained that later the deterrents fell away. Just how far this holds, and for what regions of the country, might well be argued. Richthofen saw that colonial throttling of the Mexican economy was continued by the leading families, so that many of the causes of frustration and distress remained after Independence, compounded by political insecurity and economic stagnation. In all of this chaos, banditry remained a necessity, if not a chivalrous trade.

There are many indications in our body of literature, besides those in Richthofen's explanation, that the bandits were not only, or even predominantly, common criminals or members of something like a professional underworld. They probably had their roots in communities bordering on the road and were openly tolerated, if not supported. There was something akin to "social banditry" here (Hobsbawm 1969; Blok 1972). Symbiotic relationships with politicians have been cited; accommodations with the peasantry are indicated as well. These arrangements reflect deep and generalized socio-economic desperation.

The stories told of robberies on Mexican roads seem a separate reality. Very few travellers personally experienced robbery—only two or three out of some two dozen who referred to the phenomenon. The stories, on the other hand, were relished all along the trip. There was undoubtedly real apprehension and those who knew must often have wanted to warn those who did not, but there was a pleasure in frightening others too and an undertone of mockery. The stories became part of the denigration of all things Mexican.

The romantic aspect of banditry had become stereotyped, even vulgarized, in popular European art before the beginning of the nineteenth century (Clark 1976:107). Tales of robbery from Mexico helped to fuel that romanticism. Madame Calderón de la Barca recalled Salvatore Rosa (1615–73), the influential painter of dramatic landscapes—with *banditti*—and proponent of the picturesque (Calderón de la Barca, Fisher ed. 1966:69; Hayward Gallery 1973:5; Clark 1976:107). She felt he might have travelled to advantage in Mexico, together with Hogarth, who would have been able to do justice to the ridiculous. And a good deal of what was recounted on the subject of robbery was ridiculous: the disreputable guards, the robbers who maintained their gallantry while they stripped the ladies of their jewels, and the armed travellers who mistook the blinking of the fireflies for bandit signals and were convinced that several vague shapes coming out of the night were the bandits themselves, when in fact they turned out to be cows (Mayer 1844:14).

ARMING FOR TRAVEL

Talking about robbery was one thing, but how should one actually deal with it? The Mexican wisdom of the road tended toward adaptation. The traveller was best off to carry neither too much money nor too little. If the bandits were not able to get a certain minimum, and Mexican travellers seemed to have some fairly clear ideas of what that minimum should be, there would probably be physical mistreatment. Also, it was better not to carry arms. When a resolute English traveller, carrying "a double-barrel rifle, a ditto carbine, two brace of pistols and a blunderbus," entered a stagecoach one day in Jalapa, his fellow passengers were aghast (Ruxton 1855:40). Resistance by one passenger could mean death for all. Furthermore, a successful shootout would invite revenge on an unsuspecting coachload the next day, or on the first passengers who looked like, or seemed to speak the same language as, those who had defended themselves or been abusive.

The foreign travellers' advice to each other ran in quite the opposite direction. The only way to counter the bandit was to arm oneself well and to show the arms prominently. Since bandits and escorts were undoubtedly cowards, it would probably be protection enough just to let a few barrels protrude from the coach window or to draw back one's coat to show the pistols at the belt. And should it come to actual shooting, one resolute foreigner with good weapons could expect to dispatch a good number of brigands before he sustained an injury himself. Such truculence edged with a sense of sport seems to have been the predominant attitude toward the hazards of this road.

UNDERWAY AGAIN AT LAST

Mayer, who was secretary to the U.S. legation, expressed the attitude very well as he recorded his departure from Veracruz. After he and his companions had boarded the coach, he took an inventory of their arms. Between them they had four guns, six pistols, and an old-fashioned dress sword ground to a very sharp point: "We sallied at mid-night from the courtyard, as resolved as any men who ever went on a feudal foray, to kill the first ill-looking miscreant who poked a hostile nose in our coach windows" (Mayer 1844:9).

Coach drivers always tried to leave the terminal at an impressive speed. As Mayer and his group rattled through the streets of Veracruz, it seems that they realized the armament in the coach constituted a danger to themselves in view of the rough road ahead. There was a hurried rearrangement of barrels so that if a gun should go off accidentally it would shoot out the window.

They arrived at the northwest gate, the Puerta de México, and had their documents checked. One needed to show a passport, a *carta de seguridad*, the antecedent of the modern Mexican tourist card, and an arms permit. All that in order, they left the town. It was dark and raining; a *norte* was raising a fine surf, which they could hear but not see. By morning they would be through the dunes and on to what remained of the famous road.

CHAPTER SIX

Vegetable Drapery

Just west of the dunes the travellers found the tropical luxuriance they had expected and indeed had already seen from out at sea as a green band on the coastal landscape, above the thin line of white dunes. It was to begin to compensate them for the desolate foreground of the ship-board view, the macabre town and the stifling dunes. Here they expected some distraction from the rigours of travel, as Heller put it, and reanimation (1853:40).

The entry was quite abrupt. Various travellers recall how they were suddenly hemmed in by greenery. Enthusiasm was general: what profusion! Not a tree or bush or herb here that would not have been the pride of a European botanical garden (Koppe 1835:178).

The travellers had begun their confirmation of the nature of the lowlanders in their scrutiny of the faces of the pilots that took them into the harbour and the people in the small craft going back and forth between the town and the fort. They continued to observe and evaluate them in earnest once they were in the town and whenever they halted along their journey inland. Some of the keener travellers had begun to respond to tropical vegetation while still in Veracruz, noticing what grew in the gardens and along the *alameda*. Everyone would notice luxuriance again around Jalapa, but it was here in the humid lowlands that basic aspects of their predisposition toward the tropical environment would be clearest.

Enthusiasm was general, yes, but it was usually qualified; one could not just stop and go off for a refreshing walk into this forest. The "vegetable drapery" (Tudor 1834:170) outside of the carriage window presented an infinity of intriguing forms and shades, but there was also something sinister about it. And furthermore, the forest proper was noticed to be interpersed with swamps, the nurseries of this coast's worst evil. Only later in the vicinity of Jalapa, where plants of temperate

origin were intermingled with tropical species, where the heat and
humidity eased off, and, most important, where there was no further
danger of infection with yellow fever would the observers become unreservedly euphoric.

These observations, such as they are, do let us see the humid lowland
landscape before it was affected by the massive in-migration, clearing,
and drainage of the twentieth century. Herein lies some of the factual as
distinct from the perceptual yield of the accounts. The forest cover was
clearly much denser than it is today; in fact, some travellers perceived it
as unbroken. However, the region already had been marked by man
much more profoundly than the travellers realized. This is very clear
from a view Rugendas gives us of the humid lowlands away from the
road, quite possibly somewhere in the basin of the San Juan (Figure 9).
Trees had been felled and fires set in aid of rudimentary ranching since
the early years of the colonial period. The palms he shows are indica-
tors of long human perturbations. The initial forest cover had been
reduced to isolated groves and scattered specimens of successional species.

In 1977 it was found that with only few exceptions the wetlands of this
lowland landscape, the very breeding places of miasmas, had been used
for intensive agriculture in Prehispanic times (Siemens 1980). It had

9 A view of the humid lowlands, by Rugendas in Sartorius (1961:180)

long been common knowledge that there were mounds, that is the remains of settlements, wherever one cared to look in Central Veracruz. Thus, what was virtual wilderness to nineteenth-century travellers had been a densely occupied and agriculturally productive region in Prehispanic times.

The evaluation of the coastal lowland environment that may be inferred from Prehispanic remains sets that of the nineteenth-century observers into relief. Views of this environment expressed in recent decades by those who have been involved in and those who have analysed the "development" of tropical lowlands, show the persistence of the nineteenth-century evaluation. A recognition of ancient alternatives to further deforestation and swamp drainage, to pell-mell modernization and mechanization, may well be critical to the future well-being of the lowlanders. One is led to think along these lines repeatedly as one follows the nineteenth-century travellers or covers the shortened distance now between Veracruz and Jalapa, but most particularly within the narrow band between the dunes and the hill land to the east, a landscape that we now know is both incredibly complex physically and layered over with the remains of successive incursions.

THE LOWLAND ENVIRONMENT

A description of this environment in current terms should help in the identification of the objectives of the travellers' imagery and enhance the effect of their intimations of micro-environmental and micro-altitudinal zonation. It also enhances one's respect for the acuity of some of the observers, especially the naturalists.

Most travellers flattened the topography of the terrain they were entering, as travellers still do. In fact, shallow saucer-like depressions from ten to fifteen metres above sea level alternate with undulating hills that are perhaps ten or twenty metres higher—outliers of the sloping, deeply incised plateau surface to the west. This gives the band of humid lowlands, so easily traced in outline, their internal complexity.

Climatically, this part of the journey was and is usually considered of a piece with everything from the coast to the first oaks: hot and humid. However, there are some interesting variations. Average annual precipitation just behind Veracruz is not far from what prevails over the Gulf lowlands generally, but westward the amount drops rapidly—from more than 1,500 mm near the coast to less than 1,000 mm some ten or twenty kilometres inland (García 1970; Figure 4). The volcanic mountains of the Tuxtla region to the east and the spur of the Sierra Madre Oriental that arcs toward the ocean to the north impede moisture-bearing air masses in summer and winter respectively. However, the humid lowlands, particularly their depressions, are not entirely dependent on local precipitation. In the dry season, their luxuriant vegetation is set off strongly from the dryer and fairly desolate-looking hill land. One sees what seems to be a chain of oases. The travellers usually passed through during the dry season; small wonder that they were impressed by the greenery that set in abruptly after Santa Fé.

As important as precipitation for the humidity of this zone is the impedance of its drainage by the belt of dunes. This has given the main through-flowing rivers of Central Veracruz (the Cotaxtla, the Jamapa, the Antigua, and the Actopan) sinuous lower courses and induced them to aggrade their flood plains. The seasonal fluctuation in their levels reflects variations in precipitation within catchment basins that include extensively settled mountainous terrain under a pronounced wet and dry climatic regime. They are placid and unimpressive between their high banks in the dry season but rise quickly with the onset of the rains. The mean annual fluctuation in stream level measured at various stations along the Antigua and Cotaxtla rivers, for example, is around five metres (Secretaría de Recursos Hidráulicos 1971). Every so often one or another of them rises to extraordinary levels and floods its surroundings. These can be catastrophic floods, sweeping away even the normally safe dwellings and crops on the tops of levees, marooning cattle that could not be driven to higher ground in time.

The intermittent streams draining the wetlands within the humid lowlands are sometimes referred to as *arroyos*. They join the main streams just before these exit through the dunes and are ill-defined and sluggish. The seasonal fluctuations in their level are less than half that of the through-flowing streams. This reflects the storage capacity of swampy terrain and its dampening effect on seasonal variations in run-off. Nevertheless, the seasonal rise, triggered by heavy rains is sufficient most years to flood the wetlands proper. This is relatively benign flooding; it can be adjusted to and indeed exploited.

The annual rise of water in the wetlands gives the agriculture and ranching in the humid lowlands of today a basic pulse—increasingly modified by drainage and irrigation works, but still evident. Activities move upslope with the onset of the rains and downslope in the dryer months. The pulse is barely noticeable in most travellers' accounts, which is not strange since few travellers saw this region at more than one season. One of them, coming along the road in August, did notice cattle being driven along the road; this was very likely a movement out of the flooded wetlands into neighbouring hill land after the rainy season was well under way (Ruxton 1855:32-3).

Chappe d'Auteroche, the French astronomer making his way in 1769 to the peninsula of Baja California to observe the passage of Venus, noted that when his party left the coast just north of Veracruz and turned inland, they passed through "immense woods" (Chappe d'Auteroche 1973:27). The vegetation, that many observers described without making many distinctions, may be given some detail if it is visualized as arranged along a micro-altitudinal continuum. The hills that endlessly subdivide the humid lowlands were covered, before most of them were cleared for pasture, by a low, tangled, and often thorny forest that drops most of its leaves during the dry season. With the rains the grey look of dormancy changes to a range of delicate greens. Movement downslope into the zone of seasonal inundation brings one into forest that is higher and mostly green during the dry season; only remnants of this are visible today. In the wetlands proper there is likely to be higher forest still and then hydrophytes in the depressions that remain saturated all year. The vestiges of the ancient canals that separated Prehispanic planting platforms may have open water in them during the driest months. Most of this land, too, has been cleared, and laced with modern drainage canals that lead the annual floodwaters away quickly. Plantation crops, pasture grasses, and only remnants of the former climax communities or their succession species prevail throughout the landscape that the nineteenth-century visitors perceived as fairly solidly forested.

SEGMENTS DESCRIBED IN DETAIL

A few nineteenth-century observers paid special attention to certain segments of the humid lowlands. They found them particularly evocative, as writers and artists dealing with tropical subjects have found to this day.

Wetlands

Just west of the dunes the road to Jalapa plunged, as several travellers perceived it, into a tunnel through the forest. From the grade of the railway begun in 1850, on the other hand, it was possible to get something of a look around. Several authors coming along it as the period of our concern was fading, described a wetland that lay just to the north.

One description is by Domenech, press secretary of Emperor Maximilian, who could be funny, sarcastic, and on occasion surreal (1922:31–2). He saw a cradle of death adorned with exuberant life. On islands and along the perimeter were elegant trees, woven together and strangely, permanently green. Lizards, monstrous toads, serpents, alligators, and all manner of other amphibious horrors bred freely in the mire under the beautiful aquatic plants. From between these lovely leaves emanated the evil exhalations responsible for yellow fever. Butterflies and birds filled the air; at one spot flowers shielded a rotting reptile. A traveller should pass by all this, he thought, with the speed of an express train, but of course there was no such train, in fact only a light, open railway car pulled by mules. Other observers, from Humboldt onward, went further. They were convinced that the only thing to do as soon as it was economically feasible was to drain these sources of contamination and impediments to rational land use.

Müller, the naturalist, found a journey that he made on a similar hybrid conveyance along the same route immensely stimulating (Müller 1864:203–4). He seems to have ignored the threat of yellow fever even though he was travelling inland in August. The swamp that Domenech found threatening Müller enjoyed. As birdwatchers will, he enumerated species with whatever common or Latin names he had to hand. First on his list was the jacana, a brown bird with long toes that walks on the floating leaves of aquatic plants; then the coloured species: cardinals, tanagers and orioles. He noticed herons of all sorts and finally some screeching parakeets. And with that his party arrived at the end of rail and it was on to other things. This short entry, about half the length of Domenech's fulmination, brings to mind what was being done by naturalists in Amazonia at the time. Indeed, Müller would not have been out of place in a latter-day society for the preservation of swampland. Domenech represents the view that dominated among our observers. Among the many drawings made of Mexican landscapes by nineteenth-century travellers there is a close-up of swamp vegetation (Figure 10). The artist seemed to find it both ominous and fascinating.

Recent investigations have shown that the wetland described by these two observers and many other wetlands in the humid lowlands of Central Veracruz are patterned, almost without exception, by the remains of Prehispanic canals and planting platforms, commonly referred to as raised fields (Siemens 1980). In fact, it has become apparent that the very swamp Müller and Domenech skirted on the new railway was once a garden (Figure 11).

It is postulated that the patterning represents a securing or an elaboration of fugitive use of wetland margins during the dry season of each year, when floodwaters recede and expose alluvial soils that can be exploited to advantage with rapidly maturing crops. The canalization and field build-up that may be deduced from the vestiges would have speeded the descent of the water table at the end of the wet season and access to the planting surfaces. Simple dams at the main collectors could have detained sufficient water in the system to facilitate scoop irrigation later in the dry season and to allow circulation in canoes. Thus the swamp's rhythm was harnessed. Crops could be produced within it

10 Swamp vegetation, by Pharamond Blanchard, as in *Artes de Mexico*
(166[1960]:84)

during the dry season, and if the platforms had been built up sufficiently, all year round, as in the *chinampas* still operative in central
Mexico. Together with wet season crops produced on terra firma, in
places by means of terracing and irrigation, this would provide food
products at fairly closely spaced intervals all year. To postulate such
cropping, and to add the likely ancillary activities such as hunting,
gathering, and fishing, lends credence to the reputation coastal Veracruz had before Contact as a region of abundance (Durán 1964:133–4).

The obvious readiness of the Prehispanic inhabitants to move into
and use the swamps contrasts dramatically with the nineteenth-century
traveller's unease about the wetlands of coastal Veracruz, and indeed

11 Route of the early railway, a wetland behind it with the vestiges of a complex of Prehispanic raised fields

the entire belt of humid lowlands. The ancient agriculturalists seem to have had more in common with Müller than with Domenech. Indeed, they are remarkably in tune with modern conservationists, a convergence that will be referred to again in the final chapter. Fragments of iconographic evidence from the Central Veracruzan region, as from the Maya lowlands, hint that nature in the wetlands was considered beneficent, that floods for example, were gifts of the gods, and indeed that the wetlands in general had the reputation of productivity (Medellín Zeníl 1979; Puleston 1977).

Unease was rooted in the very languages of the nineteenth-century observers. In Germanic languages the words used to designate wetland (swamp, *Sumpf*) had their origin in medieval North-Central Europe (Grimm 1942:1079–80). The corresponding words, in particular *pantano*, entered the Romance languages from Italy and are considered pre-Roman. "Swampy" terrain, of course, was extensive in both areas, and the struggle to drain it a protracted one. In both linguistic regions the various nouns applied to wetlands often become adjectives that describe a situation that is corrupt, that does not provide a sure footing (*sumpfig; marécageux*), or worse, verbs that indicate a nightmarish paralysis (to swamp or mire; *empantanar*). Since medieval times, therefore, it has been heroic to drain, to alter permanently the hydrology of a region, to firm it up, rid it of evil vapours and make it useful.

There seem to have been good reasons for the development of such an antipathy in Europe. It is apparent that malaria had been a scourge around various Mediterranean wetlands long before the explorational forays into the Americas. Malaria and yellow fever, along with the deprecating wetland terminology, were introduced into the New World with the Conquest.

Riverbanks

Koppe described a riverbank environment around the notorious little town of Medellín on the Jamapa river. By this time he had spent several years in Mexico as Consul of Prussia, had observed carefully and accumulated a fund of information that would be very useful to German businessmen prepared to operate in Mexico. He was now on his way home, and intended to embark on a passenger boat. However, Veracruz, under the leadership of Santa Anna, was in rebellion against the central government, whose forces had encircled it, disrupting normal shipping. He eventually did get on board his boat, but there were some days to spend waiting in Medellín before a way could be found to get around the besieged port and out to the anchored ship. During those days he made several excursions into the surroundings. Koppe was curt and a bit condescending about goings-on in Medellín. Wild women and gambling did not interest him much, but he did get enthusiastic about what he saw in the forest. He showed again that he was indeed an observant traveller. He made it sound as though this was the beginning and not the end of his time in Mexico.

Koppe was impressed more than anything else by the scale of this forest and ran through a string of adjectives in trying to do justice to the various trees: "gigantic," "colossal," even "heroic." Though the terrain was flat, he noted, the varied heights and massive crowns made it appear mountainous. To the Berliners among his readers he explained that if one of these trees were placed in the midst of their Lustgarten, it would overshadow all of it and rise above the height of the King's palace nearby (1837, II:316–17).

Koppe had maintained on his initial journey from Veracruz into the interior that he was no botanist, and that it was a good thing too, since otherwise he would have lost himself in the forests of the humid lowlands (1955:85). But here on the riverbank, near the end of his time in Mexico, he was ready to indulge his naturalist's tendencies and to identify tree species: "*Tamarindus indica, Dracaena drago, Cassia fistularia,*

Styrax officinale, Liriodendron tulipiferum, Bombax pertaedron, Carolinia insignis, Cocos nucifera" (Koppe 1837, II:316). Few of these names are cited in the best current material on the high tropical forest of the Mexican lowlands (Flores Mata et al. 1971; Rzedowski 1978). It might take an historically inclined taxonomist to decipher them, but that does not matter much because the list is really a poem. It is continuous with the wonder expressed by this sensitive observer at the web of blossoming, spicy-smelling plants and clinging vines in natural arbours that shaded his outings, the birds of many kinds that enlivened the trees, the wild pineapple, and the milk of coconuts there to be taken, freely, for refreshment.

Carl Nebel, a German artist who travelled in Mexico between 1829 and 1834, left many representations of Mexican customs and costumes, sketches of archeological monuments, views urban and rural. He also did a striking engraving of the environs of a lowland river crossing: it has been printed here in negative for emphasis (1839; Figure 12). The river is not named, but a location along the Jamapa is as logical as any. In the preface to the collection Nebel pointed out, as many of our authors had done, that he was dedicated to facts. He meant to inform, and just to make sure, he appended a lengthy descriptive caption to the print of the river bank. In fact, like Koppe, he aggrandized this forest, making it truly magnificent. The drawing is carefully done. The banks of the the the river, for example, imply the seasonal amplitude in river level indicated by modern measurements, and the level of the water is characteristic of the dry season, which is when travellers preferred to come to Mexico. Difficult river crossings of the kind portrayed are described in various accounts. The range of species shown is plausible, both generally and specifically, but the typical tangle has been minimized. Nebel freed the large trees, individualizing them. Using huts and a rider for scale, he made them gigantic. The lighting was arranged as in a theatre, the bamboo was made to arch, as bamboo should. The tropical forest on a riverbank, like highway robbery, and indeed, Domenech's swamp, was being made a vehicle for the flight of a romantic fantasy.

Rivers, of course, have long provided the main routes for entry into tropical lowland generally. Their banks may well have been the earliest loci of settlement and agriculture (Harris 1972:187). Riverbanks throughout coastal Veracruz are dotted with the remains of monumental architecture from a millenium or more before contact. Current settlement and agriculture are often concentrated along these same levees. Most travellers into tropical lowlands still see only what is immediately alongside tropical streams or, in recent years, roads. This is "jungle," the popular undiscriminating term for tropical lowland nature. It is actually an interesting word: it orginated in India and came to be used

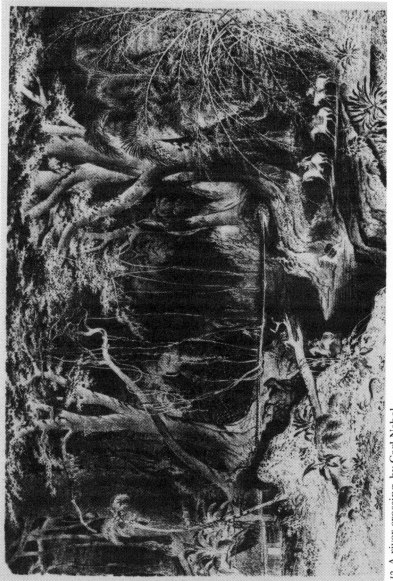

12 A river crossing, by Carl Nebel

for uncultivated ground overgrown with a tangle of vegetation. It is not particularly appropriate for high humid tropical forest proper, since that is relatively open at the base, but it is very appropriate for the tangle of vegetation that occurs when sunlight is given access to the full profile of such forest around a treefall, along a river's edge or a road. This is always difficult to penetrate; it easily becomes mysterious and menacing.

H.M. Tomlinson some years ago manipulated this imagery well in *The Sea and the Jungle*, when he described the forested banks of the Madeira sliding by the railings of his steamer: "This forest, an intrusive presence which is with us morning, noon and night, voiceless, or making such sounds as we know are not for our ears, now shadows us, the prescience of destiny, as though an eyeless mask sat at table with us, a being which could tell us what we would know, but though it stays, makes no sign" (Tomlinson 1912:178).

Several films have used riverbank imagery very well. In *Apocalypse Now* incredible tension builds aboard a boat winding into an Asian interior; the jungle is expected any moment to come alive with gunfire. In *Fitzcarraldo* a wall of Amazonian riverbank vegetation glides away behind the crazy victrola playing civilizing arias to whatever may be listening, exorcising the jungle's menace.

A Transect

Sartorius described what he saw between La Antigua and Paso de Ovejas (1858/1961:4–8, 179–81; Figure 1). It is the typical offering of a naturalist of the time: landscape data, with as much Latin as possible, pauses for pictures in words, and an occasional respectful bow to the Creator.

Sartorius was criticized for having perpetrated too favourable a picture of Mexico and of the lowlands in particular (Ratzel 1969:382). He was interested in promoting emigration from German states and undoubtedly wanted to put his whole Mexican venture into the best possible light. However, the focus of his interests was the midslope region. His hacienda, El Mirador, lay at about 1,000 metres above sea level, roughly equidistant between Jalapa and Orizaba. It seems to have been mainly in this area that he foresaw good possibilities for colonization, and hence his descriptions of other areas may not have been as affected by his biases. Sartorius cannot have had an alternative to the miasmatic theory, yet he travelled through a patchwork of wetlands and terra firma, exposing himself at length to dangers that others fled from. He discusses various natural hazards and annoyances, but he does not show the fundamental unease that many others showed. He probably felt well acclimatized and thus fairly safe from the fever, as well as knowledgable enough regarding the various other unpleasant possibilities of this environment to avoid them or deal with them. Be that as it may, he showed a lively curiosity, which he shared with other naturalists who came this way, and which for him, as for them, clearly overrode other considerations.

Sartorius begins his transect with an idyll: "A fine tropical picture lies before us, the tranquil sheet of water being surrounded by the most luxuriant vegetation; in the fore-ground are some huts beneath lofty trees, on the left bank of the river, in a forest of fruit-trees, the village of Antigua, whose ancient stone church is evidently one of the oldest in the country. The beautiful blue mountains of Misantla form the background" (Sartorius 1858/1961:6).

From La Antigua, Sartorius headed into the forests to the southwest of the junction of the La Antigua and San Juan rivers. He veered away from the western levee of the latter, crossing what he called a plain, in fact a segment of the extensive, low humid terrain, sluggishly drained by intermittent tributaries of the San Juan. This "plain" had a slight undulation, which can be read out of Sartorius' account of the muddy stretches on his trail and the vegetation he describes for the slightly higher interstices: a difference of one or two metres. A journey through this country will have been feasible only during the dry season; during the rains, floodwaters would have risen over much of it.

Sartorius describes a fig tree with all its appurtenances on a patch of slightly higher ground (Sartorius 1858/1961:7). It is typical for the variant of tropical forest that seems to have predominated in the area, and which is now designated *bosque tropical subcaducifolio* (Rzedowski 1978:183–5). The tree had a massive central trunk from which it had sent out long horizontal branches, which in turn had sent down perpendicular shoots that had taken root in the ground. It had become a vast arbour, with a huge domed crown. Its branches were entwined with every imaginable creeping plant. Orchids bloomed at the joints and *tillandsiae* (Spanish moss) floated in the breeze, "like a grey veil." *Bromeliae* covered the ground underneath—a fine assemblage!

Sartorius appreciated the tall palms he found in this area, "the handsomest which nature can produce" (1858/1961:7). He transformed a grove of them into a cathedral: the stems, twenty to twenty-five metres in height, became the columns, blossoms and fruit the capitals, and the fronds the vaulting. The dark forest all around made up the walls. He was filled with awe and reverence, and thoughts of God. The palm in question was probably *Roystonea dunlapiana*, a succession species to *bosque tropical subcaducifolio*, and an indicator of prolonged human disturbance of the wetland environment (Rzedowski 1978:351; Vásquez Yanes 1971:72–3). It occurs with varying density throughout the seasonally inundated parts of the humid lowlands west of the Central Veracruzan dunes. It is resistant to fire and hence its persistance will have been facilitated by burning in aid of rudimentary pasture management since contact and for cultivation before that.

Sartorius made an observation which enlarges our understanding of colonial ranching in the tropical lowlands and helps to fill a hiatus in the interpretation of human exploitation of the coastal wetland from Prehispanic times to the present. He saw herds of cattle in the forests of the San Juan basin, which he knew were often driven there by their owners, especially in winter (Sartorius 1858/1961:8). In other words, when pasture was scarce on the hill land during the dry season, the cattle would browse and graze in the wetlands. The floods of the wet season would displace them upslope again. Analogous tropical lowland transhumance has been described out of the environs of the Amazon by various authors, notably the German geographer Herbert Wilhelmy (1966).

Tylor was told about ranching in this country by a somewhat timorous cowboy who accompanied him (1861:323). Cattle flourished at every elevation, he maintained, in contrast to horses and mules, which were found from 5,000 to 8,000 feet. The bulls of the lowland were wild and extremely dangerous; cattle had to be enticed into enclosures by decoys.

Throughout the humid lowlands there are "islands" of higher ground, as pointed out. The variation in the relief between permanently wet swampland and the slightly higher seasonally exposed ground is one or two metres. The variation between all of this and the "islands" is in the order of tens of metres. These are the sites of ranch houses and corrals today, as they were in Sartorius' time. This reflects a concern for safety from inundation but, just as important, for the breezes that bring down the sensible temperatures there and discourage mosquitoes. These were also the loci of prehistoric settlement.

Sartorius visited a *ranchero* at his house on the top of what one must deduce was just such a hill and was given some intriguing information about the use of nearby low-lying humid land. The house site was up in "the pure air" (Sartorius 1858/1961:180). The land below could be used for dry season agriculture. One could grow very good crops down there, but it was uncomfortable to live there because of "the intermittent fever ... snakes, mosquitoes and garrapatas" (Sartorius 1858/1961:180).

Seasonal use of land subject to inundation can still be seen in a few locations within the belt of humid lowlands today, as near the village of El Palmar (Figure 1). It is a complex activity, carefully tuned to the seasonal oscillation of the water level. It provides a wide range of food crops and complements wet season agriculture on the higher land.

However, it remains a risky undertaking. In addition to the menace of wild and domesticated herbivores, there is the problem of varying water levels; floodwater may stay too long, making a mature crop impossible. And then there is the *norte*, the characteristic storm wind of the dry season, which can easily flatten a crop. Nevertheless, this agriculture on the wetland margins, with its ancient antecedents, suggests the potential for agricultural intensification. Canalization and field build-up to secure such cultivation could substantially enhance the wetland's productivity.

Once up out of the humid lowlands and into the hill land to the west, Sartorius looked back over the largely forested, palm-studded terrain from which he had just come. It is the view one has from the modern highway coming down into Paso de Ovejas from the west. The *bosque tropical subcaducifolio* reached up along the valley bottoms, but the interfluves were covered by a lower, fully deciduous forest, adapted to the lesser quantities of moisture available there. This was what the soldiers of the American army were to call *chaparral*. The "fine stone bridge" on the old highway at Paso de Ovejas indicated to Sartorius that he had left behind the true wilderness of the coastal tropics and returned to civilization (1961:8).

GENERALIZED CHARACTERIZATIONS

Most travellers through these lowlands were limited to what they could see along the main road between Santa Fé and the hill land to the west. Many would have sensed, as Ferry did, that "a wall of luxuriant vegetation on each side bars everywhere the entrance of man" (Ferry 1856:288). Tudor saw a wall of "vegetable drapery" (1834:170). There is a similar natural basis for this along a road as along a riverbank. Anyone could see that this drapery was rich in species. Humboldt had suggested that the aesthetic attraction of tropical vegetation, its picturesque qualities, lay mainly in its variety (*Essai sur la Géographie des Plantes*:30, as cited in Glacken 1967). Here it was, almost within arm's length:

All new to the European eye and thrown together in such fanciful confusion . . . each tree supports two or three creepers, the fruits and flowers of which bear no sort of proportion in point of size to the slender branches of the mother plant, it is not easy to distinguish them, at first sight, from the produce of the tree to which they cling. The air is quite perfumed at times with this profusion of flowers, many of which are most delicately coloured (particularly the varieties of the Convolvulus kind;) while the plumage of the birds, of which, in some places, the woods are full, is hardly less brilliant than the flowers themselves. (Ward 1828, II:183-4)

The animal world was sensed to be quite as varied as that of the plants. Ward saw deer bound across his road. The more exotic species, the "Jaguar (Mexican Tiger), and other wild animals" he inferred from the skins he noted in use in the settlements along the road (Ward 1828, II:184). Most of this world was heard more than seen: "The shrill noise of the cicadae, the chirping of crickets and grasshoppers, the chattering of parrots, the tapping of the woodpeckers on the dry trees, the cry of the apes" (Sartorius 1961:8).

Madame Calderón de la Barca gave the profusion of species and the whole luxuriant aspect of the vegetation between Santa Fé and Tolomé a genteel interpretation (Fisher ed. 1966:63-8). She and her husband were visiting Santa Anna at his favourite estate, Manga de Clavo, just northwest of Santa Fé (Figure 1). It was located just in the eastern edge of the patchwork of terra firma and wetland that makes up the basin of the sluggish San Juan. Most of the vegetation she saw around her will have been natural forest, but it seemed to her one vast garden.

Since early in the eighteenth century the fashion in European pleasure gardens had been the English garden. This was free of symmetry, complex, highly varied in the plants it enclosed and the aspects it offered. Nature was fruitful here, even rank, yet not without an ultimate order: that in the mind of the tasteful lord of the manor or the Creator himself (Immerwahr 1972:47-71). In fact, the English garden could be considered the very embodiment of eighteenth-century Romanticism (Immerwahr 1972:47). English tastes in landscape still run to much of this (Lowenthal and Prince 1965:190-6). Madame Calderón de la Barca saw it in the tropical vegetation around Manga de Clavo.

Other travellers noted that although this "garden" was undoubtedly luxuriant, it was not benign. Wilson expressed it well, after having just reached Jalapa and thought back about the tropical vegetation seen earlier in the day: "The whole animal kingdom busily laboring for the destruction of its kind. Reptiles prey upon each other; parasitic plants fix themselves upon trees and suck up the sap of their existence; and man, while he enjoys to a surfeit these bounties of nature, must watch narrowly against the venom and the poison that comes to mar his pleasure, and teach him the wholesome lesson that true happiness is found only in heaven" (Wilson 1855:51).

Sartorius specified some of the menace in tropical vegetation, more by way of practical warning than as a means of moralizing (1961:5). He emphasized a particularly vicious example: the *cowhage*, a leguminous plant with an interesting pod. Should the curious traveller reach out and touch it, a cloud of finely serrated hairs is released, penetrating the skin and causing an intolerable itch. He might have discussed various *Euphorbiaceae* that secrete toxic juices or a double menace, a bush of the *Mimosaceae* family with thorns an inch or more in length, they are hollow and harbour the fire ant (*Solenopsis invicta*), which can quickly raise extremely itchy pustules.

A profound ambiguity stretches through the travellers' commentary on the tropics and especially what was said of the humid wetlands. They were not quite wholesome; entering them was dangerous and vaguely sinful. Müller, the naturalist who had left Veracruz with great hopes for rich new scientific data and who was so taken with birdlife in the swamp just beyond the dunes, eventually lost his specimens and his notes before he could make proper use of them.

Humboldt, the well from which all of our travellers drank, had already intimated this ambiguity. In his *Essai* he had emphasized the overwhelming luxuriance of plant life in the tropical world, but also noted that it lacked the tender greens, grassy plains, and meadows of the temperate lands (cited in Glacken 1967:545), that is, the aspects that would allow a visitor from Europe or North America to feel at home. In his *Kosmos*, Humboldt had dwelt again on tropical luxuriance, affirming also that the old Northern European longing for a warm climate and exotica had become focused on the American tropics and that these tropics had stimulated European poetry and science immensely (Humboldt 1847, II:31–2). However, he subsequently alluded in his description of a tropical forest in South America to vegetative crowding (Humboldt 1852,I:216). The luxuriance had become somewhat oppressive, even for him. Darwin noted this carefully too and, of course, became preoccupied with the struggle among life forms (Eiseley 1961:183).

The German geographer, Friedrich Ratzel, expressed the extreme of a northerner's unease over tropical nature, especially the forests of the humid lowlands—a surrealistic landmark in nineteenth-century travel literature. Most of the other travellers are somewhere between Humboldt and Ratzel in this respect. Ratzel spent some time in Mexico in 1874 and 1875, after most of our main company of travellers had come and gone, and sent his impressions home in a series of Mexican travel essays to the *Kölnische Zeitung* (Ratzel 1878/1969). They have been little read; they are overshadowed in analyses of Ratzel's work by essays he sent from the United States, where he went first, before going to Mexico. Lumped together, this travel literature has been ranked highly as geographical description (Wanklyn 1961:14–15). It remains some of Ratzel's most readable material and evidently was an important part of the data base for his voluminous theoretical work, which influenced much subsequent thought regarding the relationship of man and nature, especially in geography and anthropology. Carl Sauer considered the concluding Mexican essay on nature in the tropics, which is of principal interest here, as "a worthy companion piece to Humboldt's classic" (Sauer 1971:250).

Franz Termer, the respected Mexicanist, in a foreword to the 1969 reissue of Ratzel's book, says of its last essay that seldom has the tropical rain-forest been characterized with such penetration (Ratzel 1878/ 1969:x). Perhaps the essay was not read carefully by Termer, or Sauer. Ratzel does not seem to have penetrated tropical forest in any sense, nor to have written so as to encourage others to do so. His culture-bound, antipathetic description of this forest does not compare well with the even-handed description by Sartorius, or, even more tellingly, with two roughly contemporaneous classics on the subject, those of Wallace, published in 1889, and of Bates, published in 1863. These authors were not afraid to enter the forest itself, all three endlessly curious about the multitudes of species and their interrelationships.

Ratzel's essay comes in two parts. In the first, Ratzel describes most effectively the humbler aspects of tropical nature that he found along the paths and fences just outside Acapulco. It is a botanical tour de force, with an enthusiastic tone, not very relevant here except for contrast. The second part of the essay deals with the tropical forest proper, generalized from what Ratzel saw in various Mexican lowlands. Its tone and import are very different.

Ratzel found a disturbing disorder in the tropical forest. He felt that earlier visitors, and perhaps he wished to include Humboldt here, had been far too favourably impressed with the superficial luxuriance of this sort of vegetation; he found it repelling (Ratzel 1878/1969:402–12). Many tree forms seemed presumptuous, unfinished, and most unlike the aristocratic, well-formed, full representations of freedom to be seen in a temperate forest. The vines and the parasites were somehow improper. Many plants were simply grotesque. All of it was an exaggeration, a confusing, pointless tangle. Nowhere could the observer find repose.

In the last paragraph of the book, Ratzel summed up his invidious comparison: "One can walk and live in our forests with enjoyment, whereas even in densely settled tropical areas paths through the forests must always be recleared before they can be used. Every entry is won at the cost of annoyances large and small. Bloodthirsty insects and thorny plants alone make it impossible to enjoy nature in tranquility, as one can at home. Finally, one is stopped before it all in amazement, as before the sea, not risking to come too close." (Ratzel 1878/1969: 411–12; translation by the author).

The counterplay of fascination and dread has persisted as a popular theme. The tropical fiction of Graham Greene, for example, presents an enervating tropical nature, together with the notorious seediness of those who have come from temperate zones to live within it. One enters such an environment at one's own peril, but it is fascinating for that very reason, as much as for the palms and the balmy night air. In diluted form, the dread is used to help sell holidays in tropical locations. The traveller is urged in subtle or explicit ways to take the risk. After any lecture on travel in the tropics, apprehensions about insects, snakes and intestinal wildlife are likely to come in the questions, and the more graphic the answers the better.

The very strangeness of this world, the dangers that actually attended an approach to it, continue to stimulate scholarship. A highly regarded, slim little volume on the tropical forest by the biologist Daniel Janzen illustrates this well (Janzen 1975:5, 36–7). There may be displays, marvellous and still not well understood, such as the bright flushes of coloured leaves that can occur in the crowns of the tropical forest at any time of year, as well as all sorts of intriguing reproductive strategies. But there are also the various secondary compounds produced by plants that can drug, gum up or kill a herbivore—and that can be highly toxic to man as well.

A SUMMARIZING IMAGE

Modern poster-makers and advertisers often allude to the paintings of a contemporary of Ratzel, Henri Rousseau, when they want to portray the tropics. Prints of his "jungle" pictures are still selling well. They express both the allure and the menace, and as it happens, they have their roots in the humid lowlands of nineteenth-century Veracruz. Rousseau may or may not have gone to Mexico in 1862 as a young clarinetist in the French army sent by Napolean to aid Maximilian (Rich 1946:9), but he is said at least to have talked to soldiers who participated in the foray (*Encyclopaedia Britannica* 1968:VIII:692). Indeed, his paintings seem illustrations of the tropical vegetation described west of Santa Fé by Blanchard, who came to Mexico in the earlier 1838 French expedition (Blanchard 1839:92–8, 211).

Rousseau has been called a primitive; he used a wide range of strong greens, painted his leaves in bold, stylized shapes, and hugely exaggerated his blossoms. Against this he sometimes placed lions, devouring lesser animals in a Darwinian fashion, at other times primates playing with large oranges. The most evocative, and best known, of these paintings is probably *The Snake Charmer*. It might have been set somewhere along the La Antigua River. The typical "vegetable drapery" is hung along the far bank. The foreground is a tapestry of Rousseau's marvellous foliage, a stylization of what the travellers approaching the Mexican coast had hoped would compensate them for the barren dunes of the coast and the macabre port. But there is menace, too—the luminous edges painted on long blades of grass, as if the artist himself had known sawgrass (*Cladium jamaicense*), one of the common hydrophytes in the wetlands. Snakes come down from the branches and up out of the grass. Evil emanates from the bright eyes of the nude green girl playing a flute.

CHAPTER SEVEN

Chaparral

Bayard Taylor, an American journalist, was coming down the road from Jalapa in the spring of 1850. He had been to California to report on the Gold Rush for the New York *Tribune* and was returning via Veracruz. Taylor was riding up on top of the stagecoach, next to the driver, also an American. About halfway between Jalapa and the western limits of the humid lowlands, they came on the rounded eminence of Cerro Gordo, site of the American defeat of Santa Anna's army in 1847. The bleached bones of the horses and mules that had pulled the ammunition wagons were still there. The driver explained that until recently there had been plenty of cannon balls lying around too, "but that every American, English or French traveller took one relic, till there were no more to be seen" (Taylor 1850, II:328).

The driver went on to tell about what he had seen in Jalapa on the day the battle was lost: "[The Mexicans] poured through the town that evening and the day following . . . in the wildest disorder, some mounted on donkeys, some on mules, some on foot, many of the officers without hats or swords, others wrapped in the dusty coat of a private, and all cursing, gesticulating, and actually weeping, like men crazed. They had been so confident of success that the reverse seemed almost heart breaking" (Taylor 1850, II:328). Neither the Mexican disaster nor the American victory need to be detailed here; the American soldiers' descriptions of the landscapes through which they moved are relevant.

It is perhaps sufficient to recall that the Mexican War began in what is now the southern extremity of Texas in April 1846. After several campaigns north and south of the Rio Grande the Americans moved southward by sea early in 1847 to establish a beachhead just south of Veracruz. The troops went ashore on 9 March 1847. The town was besieged, and then occupied on 22 March. On 8 April the army began its move into the interior in the direction of the capital. Ten days later it engaged the army of Santa Anna in the battle of Cerro Gordo, a textbook case of how to round an enemy's position. Several engagements in the uplands followed, and finally on 2 February 1848 the Treaty of Guadalupe Hidalgo ended it all, adding "a magnificent five hundred thousand square miles to the continental domain of the United States" (Hofstadter, Miller, and Aaron 1967:331).

The diplomats, the naturalists, and all the other travellers who came through the lowlands in the first half of the nineteenth century rode in one way or another; the soldiers marched. They also had to carry full packs, wait in the sun, search for water, and camp out, which could be very pleasant on nights when there was not a *norte* blowing. They had to move ammunition, artillery pieces, and stores beyond where mules and wagons could bring them. Worst of all, they had to expect to be fired on by hidden guerrillas and to be ordered to go in after them. This gave them a gut-level appreciation of particular subsets of the rigours of the tropical lowland environment: those of the dunes from which they besieged Veracruz and the hill land they negotiated to get at the enemy on Cerro Gordo.

These two regions presented various inconveniences, but the worst was their vegetation, which the soldiers called *chaparral*. Lieutenant Jamieson found it just behind his camp on the beach, and some of it was in bloom. He first expressed his version of the northerner's ambiguous reaction to tropical nature and then defined the term:

The odiferous smell emitted by these flowers was pleasant, and made me think, when I was walking among them, of some fairy tales I had read. But amid these flowers there run some of the most poisonous insects and reptiles that crawl upon the earth. . . . The best definition I can give of the word "*chapparel*" [sic] is this, that it is a thicket—of bushes, shrubs, thorns, and vines so closely matted together that it is almost an utter impossibility for man or beast to pass through it. This *chapparel* borders on each side of the road the greater part of the way to Jalapa" (Jamieson 1849:25-6).

Many of the soldiers referred to such a "chaparral" (Figure 13). They were either bringing southward a vegetational term they had become familiar with in more arid northern Mexico, where the first campaigns of the Mexican War were fought, and applying it to vegetation they were seeing initially in its dry phases, or they were adapting a term widely used in California and rural Mexico for low brush or scrub. On a recent map of vegetation much of the plant cover of northern Mexico is designated as "matoral xerófilo": a low woody growth of a variety of species, many of them spiny, plus cacti in places (Rzedowski 1978:155). This was

13 "Chaparral": one of the species of the *Mimosaceae* family of plants, a prominent element in hill land vegetation

more or less what the soldiers meant and applied indiscriminately to
anything that grew between the sea and the surroundings of Jalapa. On
the same map one sees a longer, lilting term, *bosque tropical caducifolio*,
over much of Central Veracruz. This is low and largely spiny forest, rich
with greens after the summer rains, but dormant, grey, and without
most of its leaves for five to eight months of the year.

The dunes and the sub-humid hill land are very sparsely described by
most of the civilian authors and almost entirely left out of their conceptualization of the tropics. Few had made excursions away from the main
route through the dunes; they were dangerous because the swamps
within them and along their inner margins generated yellow fever. The
sub-humid hill land was unspeakably dreary during the dry season
when most travellers passed through it. Both were negotiated as quickly
as possible. The soldiers had to spend most of their time in the lowlands
precisely within these two regions.

Their tangled vegetation, often suffocating and bewildering, always
difficult to penetrate, was seen to be complicitous. Some stretches of

terrain were given meaning in terms of warfare. Civilian authors had made the outline of the port into a line of gravestones and the swamps behind the dunes into the very sinks of putrefaction; the soldiers gave edge and point to a bleak vegetation through which they happened to have been commanded. There does not seem much in this, at first glance, to help in a conceptualization of the humid tropics, but when the descriptions in the memoirs are taken together with a rather remarkable essay by the one civilian author who gave more than passing mention to the terrain of the chaparral, as well as with the results of recent research on the remains of ancient agriculture in the hill land west of the wetlands, they help to stimulate reflections on what really constitutes luxuriance in the tropics.

LIMITATIONS OF THE LITERATURE

The military invaders were thoroughly preoccupied; their memoirs are overshadowed by more immediate dangers and more constraining hardships than those that affected the civilian authors. There is also a more urgent need to give an accounting. An example: "The Company's movements, which the Author essays to describe, were composed principally of inhabitants of Clermont County, and he trusts that a knowledge of their conduct while absent in Mexico, will not be unacceptable to those from among whom they went forth, at their country's call" (Jamieson 1849:v).

Moreover, the soldiers had few categories into which to fit what they saw. They had little background in the literature that had nourished most of the civilian travellers. There are few references to Humboldt, and there is little science anywhere, beyond the occasional unreflective reference to dangerous miasmas and differences in unhealthfulness from place to place. There was certainly very little botanical knowledge. One officer at least expressed the need: "I regretted very much while in Mexico, that I was not a botanist. The numerous plants, so new and strange to me, possessed great attractions, and I was, with others, often led astray with regard to their names and peculiarities" (Brackett 1854:68).

The eastern metaphor, the allusion to Mediterranean images, is seldom used, and there is little evidence of the transfer of the Black Legend. The soldiers were hostile, of course, having been well enough informed of Mexican perfidy and the United States' great destiny to be motivated to do battle, but their aversion is seldom expressed in the acidic terms used by American diplomats, or indeed, some of the European travellers.

There are in fact few Mexicans in the landscapes the soldiers describe; they were populated mainly by comrades. Officers ruled this world, some beneficent and some tyrannical, some just feisty, like General Twiggs, who was not averse to kicking the backsides of troops when they broke into *aguardiente* shops. Some soldiers heard screams over the noise of the bombardment of Veracruz. John Kenly, the articulate Maryland volunteer, went into the town after the siege was lifted. At one point he noted some women cleaning up a house gutted by a shell. After picking up a fragment of it for a souvenir, he realized the women were looking at him as though he were a demon: "I judged from their looks that they supposed I was glorying at seeing the effect of our fire upon their homes. How much they were mistaken the Good Father knows" (Kenly 1873:268).

The inhabitants of the settlements along the main road are implied more than described. Their stores and gardens were appropriated; it seems the American soldiers decimated the villages they passed through. The Mexican soldiers encountered in battle formations were held in contempt. They could not shoot and had little staying power. The guerrillas were another matter: they were treacherous. After the battle of Cerro Gordo had been won and Jalapa occupied, some officers reported the inevitable pretty women behind the barred windows and amazingly civil hosts—knowledgable people who seemed to welcome the American initiative. One wonders how much they were missing there.

The soldiers found a few landscape views that were beautiful enough to send home as verbal postcards: the peak of Orizaba, the gorge along the La Antigua River at Puente Nacional, Veracruz by moonlight. Even Private Ballentine, one of the most observant and engaging of the military writers, described places by formula: "The city of Veracruz is very well built, the houses being of stone, and the walls of the most substantial thickness, an excellent thing in a warm climate. The streets are wide and well paved, and its general appearance is that of a clean, neat and compactly built city. It contains a number of very handsome churches, the painted and gilt domes of which give a highly imposing effect to the view of it from a short distance" (Ballentine in McWhiney and McWhiney 1969:118). This particular statement is interesting on another count. It straightforwardly accepts a clean and neat Veracruz, the city that most of the civilian authors could not credit. Instead, they made it macabre; their predispositions were clearly a good deal more complicated.

As one might expect, therefore, the soldiers' descriptions of their surroundings were generally sparse and without much nuance. They became eloquent occasionally and briefly—when they described life in the camps, the rigours of marches, the battle of Cerro Gordo, and the chaparral.

THE DUNES

The invading force that landed on the beach opposite Isla Sacrificios in early March 1847 had to proceed directly into the sandhills in order to lay out camps and batteries around Veracruz. Equipment, ammunition, and provisions had to be carried in by the men on their backs, guns at the ready at all times since they were within range of the city's defenders and Mexican units coming to its aid. Some soldiers died from the exertion (Ballentine in McWhiney and McWhiney 1969:113, 117). Getting the supply wagons through the belt of dunes was one of the constant problems the Americans had to contend with throughout the Central Mexican campaigns. After Veracruz had capitulated and the main army had moved inland, some units remained encamped among the dunes. Reinforcements usually spent time there too before going inland. Everyone, therefore, had their stories to tell about this environment.

A curious matter-of-factness surrounds the soldiers' accounts of manoeuvring and making do among the dunes, especially in one respect. The troops had great difficulties obtaining water. They sometimes scooped it out of the marshes formed in depressions in front of dunes (Figure 2); at other times they brought it up from holes dug into the sand. "They mixed wine, brandy, lemon juice, or coffee with it, to neutralize its gall-bitter taste" (Lowe, cited in Smith and Judah 1968:380). Yellow fever was feared by soldiers arriving in the months after the first assault, that is, well into the summer and the usual season of the disease. General Scott had hurried the invasion in order to get most of the men out of the danger zone before April was too far along. Santa Anna, moreover, had counted on the fear of yellow fever to drive the Americans up in his direction, encamped as he was until just before the battle of Cerro Gordo in the region of the first oaks, safe from the disease. However, the fear was a vague one; the soldiers were evidently not much aware of the specific indictment of swamps in the origin of yellow fever. They did not fear the stagnant water among the dunes as many of the civilian writers did. Perhaps it was just that thirst was a remarkable filter, as one of them pointed out (Ballentine in McWhiney and McWhiney 1969:114).

In the sandhills the soldiers also experienced the *nortes*, and for them these were no entertainments, as they were for travellers watching at the watergate of Veracruz. Tents were torn up and sent flying, the camp-grounds flooded: "The men of a picket on duty one stormy night found it impossible to face the wind, and the drifting sand would have blinded them, had their eyes not been protected; they wrapped the capes of their overcoats about their heads, lay down on their faces, and when the weight of sand became uncomfortable, raised up, threw it off, and lay down again" (Wilcox 1892:259). The *nortes* also diminished the army's supplies; many horses, mules, and materials needed for the invasion at Veracruz and for subsequent support were lost in shipwrecks.

There were other problems: diarrhoea was widespread, for long periods it was impossible to wash clothes, and always there were the insects. Two desperate officers "slept in canvas bags drawn tight about their necks, having previously greased themselves all over with salt pork" (e.g., Maury 1894:34). "We had one relief," a Maryland volunteer reported, "one source of great enjoyment: at night we rolled among the breakers of the Gulf without danger from undertow, and this luxury of a bath strengthened us to bear the heat of the day" (Kenly 1873:288–9).

During this siege of Veracruz the troops had to read the topography of the dunes carefully. Although the shelling by the defenders of the town was not accurate, one did have to seek a minimum of cover. It was soon recognized as well that the dunes were not just barren, moving sand, but that many were fixed. It was on these that the soldiers encountered *chaparral*, both as a light surface cover and in thickets: "All the shrubs and trees of the dense chaparral bear clusters of thorns, sharp as the stings of bees, and as stubborn as bayonets. The various tribes of the cactus nation, with their innumerable needles—trifles in comparison to the thorns before mentioned—fill up the intervals between the thorn-bearing trees, rendering the whole a complete series of impregnable natural defenses" (Ballentine in McWhiney and McWhiney 1969:114–15). Another soldier noted that Veracruz "was much better defended by the prickly pear, which grew outside in an impenetrable jungle, than by the walls. Infantry could not, or would not, have forced their way through in some places at the time I examined the defenses" (Kenly 1873:270).

The full malice of this landscape bore down on the invading army as soon as they began to move inland from their camps around Veracruz: "A great many of the men, myself among the members, were ill with diarrhoea; but being of opinion that small chance of renewed health awaited those who stayed behind in the hospitals of Vera Cruz, we were all glad to get away from it; trusting for a renewal of our exhausted vigour to the purer air of the mountains, which a few days' march would enable us to breathe" (Ballentine in McWhiney and McWhiney 1969:124). In the meantime men sank to their ankles at each step. The vegetation on the dunes that were no longer mobile pressed in on them from both sides. "The heat in this chaparral I lack language to describe; it radiated from the sands and danced about in front of you, impalpable but visible, like hideous phantoms of a diseased brain" (Kenly 1873: 326). From the numerous drawings accompanying the military accounts, one gains a great sympathy for the soldiers, buttoned to the chin as they often were in heavy dark uniforms. But even they were contemptuous on occasion of "half-naked Mexicans" (Smith and Judah 1968:307).

During the first units' departure from the coast, the image of the officers in pork fat and canvas bags was matched by another, which became quite as notorious. The soldiers were throwing away all of their personal effects that did not seem absolutely essential, and soon the roadside was littered with clothing, entire knapsacks, and even cartridge braces. "I am of the opinion," one officer wrote, "that the Mexicans made a fine haul after we passed along the road" (Brackett 1854:54). This was evidently a continuing problem as U.S. positions in the interior were reinforced and supplied. A contingent marching inland might well find barrels and sacks of food staples that had been thrown from earlier supply wagons in order to make the going easier through the sand. Columns lengthened under these circumstances. Stragglers collapsed under bushes and did not catch up with their units for days, if at all. Guerrilla bands operated in the surroundings of the road from the beginning to the end of the intervention, and they dealt summarily with stragglers.

For the military, as for the civilian travellers, there was only one bit of notable scenery along the road up the long slope toward Jalapa, and that was at Puente Nacional. To Mayer the setting of the bridge was "like those of some of the romantic ravines of Italy, where the remains of architecture and the luxuriant products of the soil are blent in wild and romantic beauty" (Mayer 1844:11). Ballentine, himself a Scot, tried to convince his readers that: "The precipitous banks of the river, rocky, and ornamented with tufts of flowering shrubs, shooting out from its fissures, and suggestive of broom and bracken, blue bells and heather, render the scene exceedingly like the section of a Scotch river glen" (Ballentine in McWhiney and McWhiney 1969:126). Here was a cheering, familiar element in an otherwise alien landscape: "To a Marylander it brought back Harper's Ferry; to a Pennsylvania soldier, the stone bridge over Conestoga Creek near Lancaster" (Henry 1950:279).

ENVIRONS OF CERRO GORDO

By the time the American soldiers were approaching the scene of the battle they were fully into the low dense forest that was, in their terms *chaparral* indeed. By sheer determination they penetrated it, on the sides of a *barranca* no less, in order to come around the Mexican position. Santa Anna came to grief because he had relied on his opponents' not being inclined or able to do precisely this. The rounding, as the military literature has it, became a classic case. This, and a good deal more about the American army's way of fighting, was soon rendered into respectful, bitter legend by the defeated Mexican soldiers (Rivera Cambas 1959, X:107).

Civilian travellers before and after Cerro Gordo only had to endure the sight of the dreary vegetation which the soldiers penetrated. They had no overall term for it; instead, many took as a symbol of the whole a tree they called the *mimosa*. There are some 450 species of *Mimosaceae*. It is represented not only in the sub-humid enclave, but also in both wetter and dryer regions; it ranges in form from herbs to trees. One is given the impression by some travellers that there was really little vegetation here of any kind. Humboldt, in fact, reported that the road below Jalapa passed through terrain bare of all vegetation (Humboldt 1813, Band IV:290). There is some justification for such a hiatus in a traveller's perception. Humboldt and most subsequent travellers saw this region during the dry season when most of the leaves of its trees and shrubs had fallen—a grey, lifeless picture. And, of course, they did not usually have to go off the road so they had little opportunity to test the density of this growth. Only one author noted the selective, fleeting efflorescence that brightens this vegetation during the dry season: "The view on both sides of the road was obstructed by thick woods of mimosas, bearing, besides their own flowers, an infinite variety of parasite plants of various and brilliant colours" (Poinsett 1824:21). Mostly, the sub-humid part of the lowlands seems to have been irrelevant to the conceptualization of the tropics, which was firmly nested in the luxuriance of its humid portions.

Although Santa Anna's grand battle failed miserably, his countrymen's guerrilla activity was effective. All along the Veracruz-Jalapa road, while the initial contingents were moving into place for Cerro Gordo and then after the battle when the road became a vital supply line for the American campaigns in central Mexico, guerrillas were a menace. They were ranked with yellow fever as one of the main hazards of the lowlands. The results of one guerrilla expedition, for example, that was organized by the governor of Veracruz against an American transport of specie, was put in the following terms: "A train of sixty-four wagons, with an escort of over one thousand men, had been forced by the guerrillas to spend thirteen days on the road between Veracruz and Jalapa, a trip ordinarily requiring but five; one hundred and five men had been put *hors de combat*, while over two hundred had been made sick by exposure to the sun of the *tierra caliente*" (Wilcox 1892:414).

And it was largely the chaparral that concealed the guerrillas. The *barrancas*, which deepened as one went upslope, multiplied the hiding places, and the fine agricultural land along the streams on the floors of these canyons provided the necessary support for those in hiding. The whole setting was as advantageous for the guerrillas as it was for the highwaymen in more peaceful times.

One could expect an attack from any quarter by day or by night. The need to maintain an incessant watchfulness, while dealing with the heat, the insects, and the rigours of the march, often dispirited the invaders. The Americans, spoiling for a good, clear shot at a Mexican, could hardly go in pursuit, except with heroic effort, and even then it would probably be useless. The resulting frustration is said to have provoked many retaliatory raids on roadside communities (Rivera Cambas 1959, x:172–3).

The guerrilla attacks were seen as akin to banditry. This was no fervent nationalistic partisan activity; it was ambush and robbery. Scott was instructed by Marcy, the secretary of war, that,

> The guerrilla system which has been resorted to by the Mexicans is hardly recognized as a legitimate mode of warfare, and should be met with the utmost allowable severity. . . . those who at any time have been engaged in it or who have sustained, sheltered and protected them, are much less entitled to favorable consideration than the soldiers in the ranks of the regular Mexican army. They should be seized . . . their haunts and places of rendezvous should be broken up and destroyed. (Kenly 1873:311)

Kenly summarized the American officer's disgust: "The kind of warfare waged on both sides was entirely opposite to all my feelings. This was uncongenial work to me" (Kenly 1873:318). Both of the commanding generals of the Mexican War concurred. Scott, in several communications, showed himself particularly concerned about what he considered to be barbarous killings of sick and wounded stragglers, about attacks on hospitals—all contrary to rules of war (Kenly 1873:309). Taylor put his finger on the basic problem: pack mules, available in the country surrounding his campaigns, could have served the transportation needs of his army "provided we could bring ourselves to make war as the enemy makes it. But this is probably out of the question. We have customs which neither the officers nor the soldiers will forgo, excepting in cases of extremity. Our camp equipage, so comfortable and yet so cumbrous, our rations, so full and bulky, all must be transported" (Smith and Judah 1968:362).

Most often, however, the Americans directed their disapproval at the unworthy, irregular opponents out in the *chaparral* and their illicit way of doing things. These opponents were indeed heterogeneous bands, including colourful characters who in other times would have been bandits pure and simple, as well as priests, patriots of various stripes, and army officers rallying remnants of defeated units of regulars. The governor of the state of Veracruz included in his body of patriots some paroled prisoners. Santa Anna himself, his country's most adept organizer of resistance fighters for many years, worked around Córdoba and Orizaba after the disaster at Cerro Gordo to raise a "new army of irregulars" to attack the Americans along their lifeline (Kenly 1873:310).

There was another element in all of this: the exasperated *ranchero*. He was a cattle owner on a small scale or a herder, who might well have been called a *jarocho*, too, as we will see. He was actually a skilled horseman, and commonly wore leather, thorn-resistant clothing. He was at home in the *chaparral* and encumbered only with few simple weapons: a gun of some sort, a machete, and a lariat, with which he could pull an American cavalryman out of the saddle rather neatly. He was driven to resistance by what the American army was doing to the subsistence resources of the thinly settled lowlands. Scott had been aware of this problem, insisting to superiors that his army should be enabled to pay for their supplies; if they did not they would only increase the hatred and ruin—and starve themselves, for the owners would destroy foodstuffs rather than yield them to the invading army (Smith and Judah 1968:365).

William H. Prescott was at work on his own conquest of Mexico at the time this one was under way, but he commented on the American invasion in his correspondence with various friends during 1846 and 1847. He had come to realize from the information reaching him that the Mexicans should have concentrated on guerrilla warfare, which was their real strength, and avoided pitched battles (cited in Wolcott 1925:642). It does seem curious that Santa Anna, the veteran insurgent, should not have appreciated this before allowing his forces to be rounded at Cerro Gordo, but then he was as subject to the old military predilection for the grand battle as Taylor and Scott.

ONE PERCEPTIVE CIVILIAN DESCRIPTION

Sartorius, the German observer who lived for many years in Mexico, chose for his hacienda and settlement project a site just above the terrain the soldiers emphasized and within the favoured zone dealt with in Chapter 9. His book, which was published about the same time as the early war memoirs, indicates clearly that he travelled extensively in the region below and contemplated it carefully. He emphasized vegetation, as the soldiers had done, calling it the, "region of savannahs and prairies" (Sartorius 1858/1961:8-11). He described it as a sloping plain, "rent by fearful chasms" (Figure 3). There were lively streams down below, little water on the slopes above. Basaltic rubble was strewn over volcanic soils. The vegetation was a patchwork of herbaceous cover and the low forest. Sartorius used a more elaborate terminology, as one would expect from a careful observer with the instincts of a naturalist, but the impression one obtains is the same: "The reader must not picture to himself fair lovely meadows, but rather dreary wilderness, overgrown with low, thorny mimosas, frequently varied with larger groups of trees and small forests. There is . . . nothing grand in the character of the vegetation" (1961:9). An itemization of some of the species as the author understood them follows, and then an important observation: "In the summer months, from June to October, the tropical rains call forth a lively green, thousands of cows pasture in the rich juicy grass, and afford variety to the uniformity of the landscape. With the cessation of the rains, the prairies fade, the soil dries up, the trees lose their foliage, the herds seek the forests and chasms, and in the cloudless skies, the sun scorches up the unsheltered plains" (1961:9).

Sartorius was linking two environments into one ecosystem; the aptness of this in lowland Central Veracruz has only recently become apparent. The seasonal rhythm of ranching was then and still is the most obtrusive functioning manifestation of this relationship. From the higher wet season pastures the cattle need to come down on their own or be brought down to where soil moisture is sufficient throughout the dry season to keep the vegetation green—the wetlands that nourish the forests Sartorius refers to or the banks of the entrenched streams.

Recently an altitudinal interrelationship of more limited amplitude has become apparent in Central Veracruz. Several communities near or within the humid lowlands have been cultivating the margins of wetlands in dry seasons and integrating this with agriculture on terra firma during wet seasons. Similar systems have been observed in many other tropical lowland regions. Just downslope from the Central Veracruzan wetland margins, within the swamps themselves, are vestiges of Prehispanic planting platforms and canals. This raises many interesting questions. Does present fugitive agriculture on the wetland margins represent long tradition and constitute, perhaps, an echo of the early stages of the intensification represented by the vestiges within the swamps? Why is there no knowledge of the latter among the rural people of today and might it not be useful to reintroduce it? The detailing and conceptualizing of past and present altitudinal integration of these subsistence activities is clearly a research imperative.

Sartorius concluded his description of the sloping plains with a paragraph that is one of the most intriguing in the book. He was, it is clear from the context, considering mainly terrain to the south of the Veracruz-Jalapa road, but the observations are relevant some distance northward, too. It had, he maintained, a peculiar charm for men of an enquiring mind, as if to say that none of what followed was to be expected of the soldiers!

Traces of extinct tribes are here met with, of a dense agricultural population, who had been extirpated before the Spaniards invaded the country. When the tall grass is burnt down, we can see that the whole country was formed into terraces with assistance of masonry, everywhere provision had been made against the ravages of the tropical rains; they were carried out on every slope, descending even to the steepest spots, where they are often only a few feet in width. In the flat valleys are countless remains of dams and reservoirs, mostly of large stones and clay, many of solid masonry, naturally all rent by the floods at the lowest part, and filled with earth. On the dry flat ridges the remains of large cities are found, forming for miles regular roads. The stone foundations of the houses may be recognized, covered with heaps of rubbish and stones, large squares with symmetrically arranged stately edifices, the principal front adorned with temple pyramids, from forty to fifty feet in height; there are also traces of plaster and mortar, and of pavements. There where the union of two ravines with perpendicular rocky walls (and there are many such points) forms a projection protected on three sides, are castles of solid masonry, with ramparts and battlements; in the courtyards are extensive remains of palaces, temples and graves. All is now concealed by trees or tall grass; for many miles scarcely a hut is built, where formerly every foot of land was as diligently cultivated as the banks of the Nile or the Euphrates in Solomon's time. (Sartorius 1961:10)

Contemporary publications on settlement and agriculture in Prehispanic Central Veracruz do not show such densities in the hill land between the Cotaxtla and La Antigua rivers (Sanders, *Handbook*, Vol. 11, part 2). Sartorius is not cited in them anywhere and his passage does at first seem a curiosity against that background. Donkin in his recent synthesis of material on "Agricultural Terracing in the Aboriginal New World," cites Sartorius, but was unable to find any of the remains during his own ground reconnaissance, which is not strange since they are extremely difficult to detect from the ground (Donkin 1979:53).

Sartorius' description was not a flight of fancy, a romantic rendering by an enthusiast for European immigration into Mexico. Recent air reconnaissance has revealed very extensive patterning underneath currently cultivated fields on those same seaward sloping surfaces dissected by canyons (Figure 14). The lines do seem to represent a strategy for

14 Remains of ancient field systems on the sloping hill land near Rinconada

making the soil more tractable and impeding erosion. They may be field boundaries; food crops or cotton are likely cultivars. In addition, recent site surveys by Mexican archaeologists, particularly Mario Navarette Hernández, indicate a considerable band of Prehispanic settlement in the lowest extremities of the sloping terrain and its hilly outliers immediately adjacent to the many wetlands of the humid lowlands.

Again, nineteenth-century observations seen against the Prehispanic background and with present practice in the foreground help us to identify longterm shifts in the assessment of several elements of the tropical landscape. Both wetlands and chaparral were seen in decidedly negative terms, as they still are, in large part, to this day. They were excepted from the overall impression of tropical luxuriance. That came from the tropical forest seen on riverbanks and along the road through the humid lowlands, as well as the happy combination of tropical and temperate species around Jalapa, projected downslope. Wetland and chaparral were not part of the conceptualization of that green band above the dunes, seen from the decks of arriving ships, where travellers had been led to believe was tropical nature par excellence. Sartorius intimated and recent investigation has abundantly confirmed that what was seen as interstitial and repulsive had once been attractive and sustaining.

In Sartorius' description of the remains of ancient agriculture there is amazement and respect. He claims at the outset of his book that he could aspire to do little more than embellish what Humboldt had already written (1961:vii), by which he was doing himself scant justice. Humboldt's wetlands were miasmatic, his sloping plains desolate, his lowlanders lazy, and the living in the lowlands by and large too easy. Sartorius agreed that nature in the lowlands was overbounteous and that the inhabitants he saw there, especially the *jarochos*, were little inclined to work. He told some fine stories about how they did things and so, in fact, was embellishing Humboldt, but almost in spite of himself, he was doing more. In his treatment of the whole coastal region there are intimations of a subtler view—subtler by far than what the soldiers saw. He distinguished among the various habitats within the lowlands, for one thing, and entered a very positive evaluation of what he had detected regarding ancient land use. Between that and what he could see being done in his time there remained a baffling contrast, similar to the contrasts that have occurred to anyone working in recent years on the interpretation of Prehispanic agricultural remains. Terrain that was once intensively cultivated is now used extensively, if at all.

And finally, in the pages of his book that treat this region, Sartorius intimates something more. Cattle are shown to be moving up- and downslope with the seasons, peasants living on breezy outliers of the hill land, but cultivating the humid land environments has been traditional wisdom here and in many other tropical regions. The potential for such integration, rather than the wealth of species in an expanse of "jungle," constitutes the real luxuriance of the lowland tropics.

CHAPTER EIGHT

Jarochos

Several formulations are in the background of our observers' assessment of the people of the lowlands. Humboldt had considered that the Veracruzan lowlands could be very productive, "if the number of colonists was greater, and if their laziness, the effect of the bounty of nature, and the facility of providing without effort for the most urgent wants of life, did not impede the progress of industry" (Humboldt, Black ed. 1911, Book III:253-4).

This idea did not originate with Humboldt. It has been traced into antiquity, and more particularly to Herodotus. The degree of civilization attained by peoples was seen as inversely proportional to the fertility of their soil and the luxuriance of surrounding nature—a variant of the idea that necessity is the mother of invention (Glacken 1967:547).

Some of the observers are also likely to have been influenced directly or indirectly by pronouncements of Count Buffon, the French naturalist whose monumental *Histoire naturelle, générale et particulière* had been published volume by volume in the latter half of the eighteenth century. He regarded primitive man in the New World as weak, unable to clear the forests and drain the marshes, unable to change nature to the degree necessary for a high civilization, unable to build well-ordered landscapes such as those of the Old World (Glacken 1967:680-1).

It has been argued that Humboldt and Darwin after him were vastly more interested in nature than in man, and thus deprived ethnology of valuable stimuli (Eiseley 1961:154-5). Such a disproportion is certainly evident in Humboldt's discussion of the Mexican tropical lowlands. However, that hardly seems to do him justice. In his description of the physical stature, customs, and languages of the Chaymas, in what is now northern Venezuela, for example (1943:132-42), and in the not inconsiderable other ethnographic material that resulted from the South

American journey, Humboldt showed himself open and tactful, as well as a careful and humane observer.

In his general discussion of the Indians of Mexico, Humboldt observed, "I know no race of men who appear more destitute of imagination. . . . I deliver this opinion, however, with great reserve. We ought to be infinitely circumspect in pronouncing on the moral or intellectual dispositions of nations from whom we are separated by the multiplied obstacles which result from a difference in language and a difference of manners and customs" (Humboldt, Dunn ed. 1972:57–8). This, too, reflects well on the last of the Renaissance men, and may well be the best explanation for his reticence in matters ethnological.

That Humboldt nevertheless did not question the old aspersion on tropical peoples but rather dismissed the Mexican lowlanders may have a good deal to do with the circumstances of his stay in Mexico. Time was limited; his travel from the plateau down to Veracruz was hurried and perhaps, therefore, could not yield the same deliberate and detailed observations as much of his other travel. In characterizing the Veracruzan lowlands and lowlanders he had to rely heavily on materials already available in the capital. Moreover, he himself wrote about the antipathy between uplanders and lowlanders in various parts of the New World (1813, IV:319–20). Humboldt's judgment may thus at times have been compromised by his good upland connections.

Most of those who came after Humboldt evaluated the lowlands and their inhabitants as he had done. There are various reasons for their failure to question inherited wisdom. As we have seen, the military men were hardly aware of the local people and seldom went beyond the briefest of clichés in their descriptions. Civilian authors generally spent only a few days along the lowland portion of the road. Some made excursions to one side or the other. Koppe was thus able to write well of resort life in Medellín; Ferry became aware of the customs of the *jarocho* in villages to the south of the main road. Only Sartorius lived and travelled in the lowland over a prolonged period. For the most part, the observations recorded were made at rest stops or while under way. Few travellers had the time or the knowledge of Spanish necessary for any sustained conversation. We thus see the material aspects of culture, and only the surface of a very limited range of behaviour, which is evidently apparent throughout nineteenth-century commentary on the Indians of Mexico (de Lameiras 1973:50–1). That of Tylor, who was to become the founder of cultural anthropology, cannot be excepted.

Payno, a stranger in the lowlands even though he was a Mexican, mused while riding out of Jalapa in the direction of Veracruz about the way in which one perceived the landscape out of a carriage window (in

Tavera Alfaro 1964:98). It passed by as a moving display, a breeze, leaving vague impressions. These may be diagnostic of the predisposition of the observer, but do not serve very well in the reconstruction of life and landscape in a particular region during a particular epoch.

The general unreliability of foreign travellers' remarks in questions of fact definitely limited ethnological studies in the nineteenth century (Lowie 1937), as well as present-day attempts to reconstruct the society and culture of the time. Yet, what is one to do for sources, de Lameiras has asked, if one rejects commentary that does not conform to standards that the early observers were not yet in a position to adopt? (de Lameiras 1973:50). Early ethnologists had to rely heavily on precisely this type of material. Amateurs were prominent in early anthropological societies in Europe and the United States (Conklin 1968:173). It is evident from many prefaces to accounts analysed here that the writers expected to have their observations about the customs and conditions taken seriously. Naturalists and diplomats, commercial representatives, and tourists of independent means all protested in one way or another that they had taken pains to ensure that "their information was correct, their descriptions faithful and their observations just" (Robertson 1853, 1:viii).

At least half a dozen of the travellers who followed Humboldt explicitly perpetuated the idea that the lowlanders were lazy. "Nature appears to have done everything, but perfect her work in *man*," one of them maintained. "His character is here marked by indelible though negative signs, as a lazy and unprofitable cumberer of the ground. . . . I remember not to have seen even *one solitary field*, during the distance of *seventy miles*, in a state of cultivation" (Tudor 1834:178).

NOT "'ONE SOLITARY FIELD'"?

The search for data on what was actually produced in the area traversed by the Jalapa road is frustrating. Enrique Florescano, in his review of the economic history of the early decades of the nineteenth century, refers to some little-used but promising sources. They are, in fact, difficult to resolve (Florescano 1976:435-53). In 1831 an enlightened governor of the state, Lic. Sebastián Camacho, published an *Estadística del Estado Libre y Soberano de Veracruz*, in which were included the reports of the heads of the state's four *departamentos*: Orizaba, Veracruz, Acayucam (now usually written Acayucan), and Jalapa (Pasquel, in prologue to Iglesias 1966:vii-viii). Koppe evaluated these data carefuly and produced various summarizing tables. Subsequent authors relied on the same sources or accepted Koppe's summations. Unfortunately, the areal subdivisions are not very clear; the degree of detail is uneven; critical figures from Jalapa are missing entirely. There is the potentially comparative information gathered in 1803 by the secretary of the *Consulado de Comerciantes* in Veracruz, José María Quiros, a source Humboldt seems to have consulted (Secretaría de Hacienda y Crédito Público 1944; Florescano and Gil Sánchez 1976:62–107). These various sources, therefore, are difficult to align. And in any case, such figures on production as can be isolated are likely to include only what was monetized, thus leaving out a great deal.

From Koppe's analysis, which is perhaps the most useful here, it is apparent that the agriculture in the environs of the road was mainly shifting cultivation and yielded little more than subsistence crops (Koppe 1837, 1:164–73). The quantities of cotton, sugar, or other commercial crops were small. This was a regional reflection of a national condition, as Richthofen later pointed out (1854:244–56). Mexican agriculture at mid-century could not fulfill the country's needs for food and fibre. Although cotton did very well in the tropical lowlands, for example, the budding textile industry in the central region could not be adequately supplied internally, and hence management relied more and more on imported raw material. This in turn jeopardized even further the prospects of native commercial agriculture.

Ranching was carried out on various scales; Sartorius has the best description of it (Sartorius 1961:181–6). Smallholders, either renters or owners, usually had a small complement of animals out to pasture, using them for subsistence purposes and as a source of cash in emergencies. "The animals cost nothing to feed, they increase without requiring any attention, therefore why should they not be kept?" (Sartorius 1961:182). At the other extreme were the large ranches, such as those of Santa Anna. Sartorius described ranching practise on such holdings: "The horned cattle are left entirely to nature; like the deer in a park, they seek their own pasture, keep together in herds or families, and choose favorite spots, to which they invariably return. According to the season, their instinct leads them to pasture, during the rainy season in the savannahs, during the dry months in the shady forests. These animals, however, are not wild, they do not shun man, and every head is marked" (Sartorius 1961:182). There was little control over breeding or ranging. The larger holdings seem to have been defined in approximate terms. The cultivator had to fence his fields against the cattle. Cattle were slaughtered for their hides and tallow, as well as beef, which was consumed fresh or dried in long strips, a curiosity often commented on by the travellers. Cows were seldom milked, except near towns, where there might be a market for fresh milk.

The essentials for subsistence were just too readily available throughout the territory between the sea and Jalapa. The materials for a house could be cut down anywhere. Corn, chile peppers, and beans could be obtained, apparently, with very little labour. Fruits needed only to be picked. The banana was so easy to get and so nutritious, it was a scandal. Every native dwelling had its complement of indigenous and introduced domesticated animals. Wildlife was everywhere, to be caught, hunted, and fished. Bees provided the sweetener; the *acrocomia* palm a good wine. Ready money could easily be obtained by making charcoal and selling it in Veracruz, or gathering vanilla, dye materials, spices, or drugs. "Can we then wonder," Sartorius asked, "if the natives enjoy the banquet thus prepared for them, and deem it folly to care for the future [?]" (1961:13).

Humboldt had already outlined the region's potential for commercial exploitation (Humboldt, 1811, Book III:250–4). As noted, he seemed to be echoing information that was being gathered in New Spain at the time of his arrival in 1803, in particular a report from the secretary of the consulado of Veracruz, José María Quiros: "Those who have said and actually believed that the terrain from Veracruz to Jalapa and Las Villas is unproductive, have been enormously deceived. The opposite is true" (translation by the author; in Florescano 1976:64).

Vanilla, the roots of the *Convolvulus jalapae*, from which could be derived "one of the most energetic and beneficial purgations" (Humboldt, 1811, Book III:253), spices, teas, tonics, and flavouring agents, including vanilla: all these could be gathered wild. Cacao, tobacco, cotton, and sugar cane could be cultivated to great advantage. Koppe later added a long list of useful woods that could be taken out of the forests (1837, II:159–73).

The use that had been made of all this was considered trivial. The agriculture seen along the road, even by those on horseback, consisted of small patches of corn or chile peppers around the houses, perhaps a dozen banana stalks, and some fruit trees (Thompson 1846:12; Sartorius 1961:6). There was little evidence anywhere of the use of the plow and this was considered fundamental.

DISTINCTIONS IN THE INDICTMENT

Koppe noted that in 1831 the canton of Veracruz had 24,556 people (Koppe 1837,I:150,153–4). This included the port, Alvarado, Medellín, 21 haciendas, 600 ranchos or small communities, and 149 *hatos*, or isolated ranch headquarters with perhaps a single family in residence. Of this population, 408 (1.7%) were of Spanish and Cuban origin, and 349 (1.4%) from other foreign countries. For the remainder, the vast majority, Koppe could only repeat the qualitative elaborations available to him: 23,799 were Mexicans, that is, minorities of Creoles and Indians, a small minority of blacks, and then many mixtures of the three. The figures for Jalapa and its surroundings were not available, but the categories will have been the same and the proportions little different.

The small minority of foreigners in the lowland population was explicitly exempted from the indictment of tropical lowlanders; although they had various faults of their own, they were still seen as lights in a sea of cultural darkness. The Creoles—Mexican-born whites, who were mostly urbanites, people of means and reason—were often criticized. Their women smoked! Their efforts at higher culture were seen as pathetic, and all were judged as addicted to gambling.

The blacks were seldom referred to as a separate group. One American writer did deliver himself of a rather glaring opinion: "The negro, in Mexico, as everywhere else, is looked upon as belonging to a class a little lower than the lowest—the same lazy, filthy, and vicious creatures that they inevitably become where they are not held in bondage" (Thompson 1846:6).

The Indians, on the other hand, were more often noticed. The descriptions were predominantly critical, but more condescending than outrightly disparaging. Several travellers saw them in passing, between the fruits and the vegetables in the market of Veracruz (Bullock 1866:15; Kenly 1873:368). Ruxton found them there as well, but "neither picturesque in dress nor comely in appearance. They are short in stature, with thick clumsy limbs, broad faces without any expression, and a lazy, sullen look of *insouciance*. They are, however, inoffensive people, and posess many good traits of character and disposition" (Ruxton 1855:28).

Rugendas has left an interesting drawing of Indians as one might have found them in the vicinity of the Sartorius hacienda (Figure 15). It is a somewhat disorderly scene, with various amiable groupings, with people and pigs in close proximity. Some industry is indicated, as well as just a touch of the risqué, from which few foreign observers could refrain. Around it all is the romanticized environment. The whole exemplified the meaning given in romantic parlance to "picturesque"; it is a tableau, rather like those still to be seen in the Museum of Anthropology in Mexico City. There are such vignettes of Indian life in many of the accounts, and one of the striking examples, somewhat pointed in regard to laziness and already a part of what would become an extended commentary on "primitive" people, but still a virtual companion piece to Rugendas' drawing, is found in Tylor's description of an Indian household just south of Jalapa. He found it

so different from what we are accustomed to among our peasants of Northern Europe, whose hard continuous labour is quite unknown here. For the men, an occasional pull at the *balsas* (the rafts of the ferry), a little fishing, and now and then—when they are in the

15 An Indian tableau by Rugendas in Sartorius (1961:64)

humour for it—a little digging in the garden-ground with a wooden
spade, or dibbling with a pointed stick. The women have a harder life
of it, with the eternal grinding and cooking, cotton-spinning, mat-
weaving and tending of the crowds of babies. Still it is an easy lazy
life, without much trouble for today or care for tomorrow. When the
simple occupations of the day are finished, the time does not seem to
hang heavy upon their hands. The men lie about, "thinking of nothing at all" and the women—old and young—gossip by the hour, in
obedience to the beneficent law of nature which provides that their
talk shall increase inversely in proportion to what they have to talk
about. (Tylor 1861:315)

Madame Calderón de la Barca saw her first Indian village in Santa Fé,
and perceived it as just such a composition, prepared for her benefit:
"the huts, clean and pretty, composed of bamboo and thatched with
palm leaves; the Indian women with their long black hair standing at
the doors with their little half-naked children; the mules rolling themselves on the ground according to their favourite fashion; snow-white
goats browsing amongst the palm trees. The air was so soft and balmy . . ."
(Calderón de la Barca, Fisher ed. 1966:63–4).

Bullock went to the village of Jilotepec, northeast of Jalapa, to watch
an Indian "religious fête." He saw "a simple, happy people, who were
performing a religious duty to their Creator, in a manner which to them
appeared the most acceptable" (Bullock 1971:466, 469). After the procession, "every house was a scene of merriment and feasting; some were
a little merry with pulque and a pleasant liquor prepared from the dregs
of newly distilled spirits; but none were rude—all was happiness and
pleasure" (Bullock 1971:471).

Ward observed how little the Indians required for their subsistence
and, since they had rather primitive notions of decency, how little they
needed by way of clothing too (Ward 1828, II:184–5). These points have
been made repeatedly, before and since, no doubt to justify low pay, as
Deleon has pointed out (Deleon 1983:19).

The Austrian countess in the retinue of Maximilian, Paula Kollonitz,
was moved by the momentary views out of her carriage window into
what she perceived as Indian yards, where men stood holding children,
women sat with chickens in their laps, and all was redolent of poverty
and patience. These people appeared to have few needs, certainly in
regard to clothing, and even less in regard to cleanliness (Kollonitz
1867:76).

Koppe used a kind of shorthand to characterize the Totonaca of
Misantla, a Cantón just to the north of the Veracruz-Jalapa road. They
were well-disposed, peaceful, and fun-loving, much given to music and

dance, greatly attached to their language and their localities (Koppe 1837,1:153). He added a cryptic note to the effect that he saw the extreme attachment to their homes and villages as a deterrent to civilization. Their mental horizons would remain as limited as the circuits of their daily lives.

Several authors equated Indian with low class. Thompson, altogether the frankest of our authors in these and other matters, put it simply: "All the laborers in Mexico are Indians; all the large proprietors Spaniards or of mixed blood. . . . So of the army; the higher officers are all white men, or of mixed blood, the soldiers all Indians" (Thompson 1846:12). Thompson also could not resist a crude comparison of the Mexican village with an American Indian village, which broadened the generalization further: "The same idleness, filth, and squalid poverty are apparent" (1846:12).

Sartorius has left us the most detailed description of the Indian found in our body of literature, and the severest indictment (1858/1961). He wrote of the Mexican Indian in general, and hence his description has little specific ethnographic utility. But it is apparent from the context that he was referring mainly to those Indians he had come to know during his long residence in Central Veracruz, and thus his description contributes substantially to the imagery under discussion here. What he says does sound contemptuous, as de Lameiras notes, and for this she ranks his book well below that of his main competitor, Mühlenpfordt (de Lameiras 1973:32).

Sartorius is indeed less palatable in his judgment of the Indians than in his many observations on the physical environment, the agriculture, and the ranching of the lowlands. Indians have inferior sensibilities, he notes. They do not seem to feel pain as Caucasians do. They are innately beasts of burden. They may well show diligence and perseverance, hardiness and practical sense, but they are incapable of higher intellectual activity. There is no imaginative spark, no initiative, and hence in a fundamental sense, they are indolent: "The educated Indians, and there are many who devote themselves to jurisprudence and theology, learn their respective sciences, but never get beyond their compendium. We find in them the talent of imitation and comparison, perhaps humour and wit, but no poetry" (Sartorius 1858/1961:64).

He goes on: there is little love within the Indian family. Man and woman are bound together by habit. Her lot is very much the worse; the way she bears it is often noble. Fortunately, these people have few wants, we read again. Therefore they are able to do quite well in the beneficent tropical environment. They are, however, completely improvident, incapable of saving, and easily fall into debt.

The Jarocho

A number of the travellers recognized this as a subculture within the lowlands. There are several brief assessments, and two romanticized and fictionalized accounts of life among the *jarochos* (Ferry 1856; Biart 1959). Then there is Sartorius again, who lived for many years among them and whose rich descriptions give our enquiry a special fillip. All this may be held up against Melgarejo's absorbing and quite plausible, if somewhat heterogeneous, recent volume on the *jarochos* (1979), as well as some slightly bizarre material from the Mexican publisher and author Leonardo Pasquel, editor of the defunct *Revista jarocha*. This elliptical, passionate, and sometimes less than careful booster of Veracruz can perhaps be taken seriously when it comes to the folklore of his native state, considering the obvious range of his personal connections, as well as the scope of his various compilations.

Disparagement is embedded in the etymology of the term, which provides a series of possibilities (Santamaría 1959:630; Melgarejo 1979:50–4). The word may come from the Arabic *jara*, a shrub, and by extension refer to *vara*, a rod or goad fashioned from such wood. This was used in peninsular cattle management prior to the Conquest, with a knife attached, to bring down cattle by hamstringing them. This would seem to be particularly practical where cattle ranged in brush or forest. The *arribenos*, that is, the people of the Mexican uplands, used a *reata* in range cattle management and may have derived from *vara* or *jara* a term applicable to the *costeños* who used it instead. The *jara*, with a point hardened by fire, was also used in the Veracruzan lowlands as a javelin-like projectile. A smaller version, the *jarilla*, was used as an arrow. Both are evidently still common toys. The *jarocho* has long been known as a formidable man with a weapon, not only the *jara* but the *machete* as well. The suffix -*ocho* seems once to have been a superlative, but came to be used with *jara* to connote contempt.

A second possibility is that *jarocho* comes from *jaro*, also an Arabic term, which denotes a ruddy colour. The word was used in Spain, it seems, for Ethiopians, and in the New World of the sixteenth century for people of mixed racial background. There has been the suggestion, too, in this connection, of a robust physique.

Pasquel has noted some other possibilities, and is confirmed by reputable dictionaries (Pasquel 1979:205; Wehr 1971:231; Real Academia Española 68). *Jaro* was applied in Andalucia and in New Spain to pigs, and *jarocho* to swineherd. The term was broadened idiomatically to designate someone repugnant. Also the word *xara* was an Arabic word used on the peninsula during the Reconquest to denote human excrement, thus deepening the repugnance.

All that adds up to fulsome disparagement. *Jarocho* has largely lost
these connotations. In recent folklore it seems a more neutral designation for a subculture. An association with herding and horsemanship is
common in the nineteenth-century literature. It has been used since to
designate *campesinos*, that is, peasants, and fishermen, as well as
herders. It can mean urban as well as rural people. Indeed, it is often a
synonym for Veracruzano. For the people of the state themselves it is
now a vague mark of distinction, connoting good-humoured pugnacity.
Entrepreneurs have trivialized it on folkloric album covers and a medio-
cre soft drink, *jarochito*.

Mühlenpfordt characterized *jarochos* in a brief formulation: they
spent most of their life on a horse, working with cattle, they were without
feelings, untrustworthy, and lazy (1969, II:46). Sartorius elaborated on
the laziness, but seems to have been more bemused than offended by it;
in fact he sounds almost envious.

He found them dressed in leather, spurred, and always carrying a
machete. Excellent horsemen, they could negotiate brush and forest easily, and amidst all that, throw the lasso with precision. They were "simple and hardy. . .obliging and always in good humor" (1961:10).
Unfortunately there was in all this a lesson about the *jarocho*, and his
countrymen in general: "The Mexican is fond of cattle-breeding,
because it feeds him without hard work, enables him to indulge his taste
for a Bedouin life, and to be on horseback as often and as long as he
pleases" (Sartorius 1858/1961:181). Over some of the habits of the *jarocho* one could only shake one's head:

> [he] would be ashamed to carry a *cántaro* of water on his back,
> although the river is scarcely fifty paces distance from his house, he
> ties his two large jars together, hangs them across Dapple's back,
> mounts behind and steers for the stream. Arrived there, he rides so
> far into the water, that the jars are filled of themselves, so that he has
> not even the trouble of dismounting. If fuel is wanting, the man rides
> out to seek for a dry tree, already blown down by the wind, and which
> is precisely thick enough to be conveyed by his beast. By means of a
> strap he fastens the end of the wood to the horse's tail, which must
> now drag the wood and of course carry his master besides. Arrived at
> the hut, the log is not cleft, but is passed in at the open door to the
> fire, and when the end is consumed it is gradually shoved in further,
> until at the expiration of some days, the house will hold it. This is the
> tropical *savoir faire*. (Sartorius 1961:6)

Mühlenpfordt specified that his denigrations of the *jarocho* applied
only to the men. Their women, the *jarochas*, took care of house and

field, were of a mild character, agile, industrious, and more honest than their men (Mühlenpfordt 1844, 11:46). The meanness and inadequacy many travellers attributed to Mexicans generally are also limited— explicitly in at least four accounts, and implicitly in others—to the male. The women's "kindness of heart and many sterling qualities, are an ornament to their sex, and to any nation" (Ruxton 1855:iv). Old Mexican matrons were seen as "the kindest mortals in the world" (Wilson 1855:48). Mexican camp women carried the burdens and cooked the meals of their husbands or lovers, "devoted creatures who follow [their men] through good and evil" (in Smith and Judah 1968:216). Müller noticed women carrying immense loads to market while their men walked ahead unencumbered (1864:245). Later the man would lie drunk at the side of the road, having spent on himself the money gained, while his wife waited beside him, watching over him while he slept it off. That was devotion, or perhaps just stultification.

Ratzel had an explanation (1878/1969:320–1). Women seemed to be fulfilling the tasks that nature and custom had given them better than the men fulfilled their duties because they were restricted in their responsibilities to a traditional realm hardly touched by civilization. They were thus much less likely to show the incompetence their menfolk showed in economic and political endeavours. The distinction many observers had made was therefore more apparent than real.

We are also shown *jarochas* in another light. Müller describes them on the *alameda* of Veracruz wearing captive fireflies in their hair (Müller 1864, 1:205–7). Biart saw them at the doors and windows of Alvarado during a horse race: flashing eyes, jewelled fingers, no stockings, or even shoes, cigars clamped into white shining teeth, and all in the costume of the region, which included a rather loose-fitting blouse that bared the shoulder and on occasion slipped, revealing breasts—without anyone's being scandalized! (Biart 1962:20–1). The *jarocha* shown thus as a seductress is made the cousin of the Andalucian gypsy, the slightly darker girl who might fascinate a white gentleman but was not usually married. Payno, the young Mexican traveller of good upland background, tells how he saw a group of *jarochitas* in a theatre in Veracruz (Payno in Tavera Alfaro 1964:114–15). They were beautifully costumed, very clean, dark in complexion, Andalucian in their grace, *but*, very provincial.

The travellers noticed only the most obtrusive aspects of the material culture. They commented repeatedly on the tropical lowland house of pole walls with a palm-thatched roof. They saw it out of the coach window in passing, but were also able to enter it on occasion when seeking shelter at the smaller stops or when buying provisions. The front rooms of many roadside houses, then as now, were arranged as small stores, or *tiendas*. One could buy alcoholic refreshments there: pulque and the fermented juice of sugar cane, or *aguardiente*, its distilled form. Several travellers admitted to adding a little something to the water in their canteens.

In any case, the house of the *jarocho* intrigued many observers. It emerges from the accounts like the backdrop for a museum exhibit of mannequins wearing native costumes and doing native things. "Pretty and fantastic," Ward called it (1828,II:195). Poinsett was amazed at the construction of its roof, the framework of canes, and the thatch, which did not leak even though one could see light glimmering through it (Poinsett 1824:20).

Mayer called such a house a "chicken coop" (Mayer 1844:10), as did several other American writers. Müller referred to it, perhaps a little more graciously, as a bird cage (Müller 1864:203). Its furnishings were sparse: a raised hearth and some cooking utensils, a few wood and leather stools, perhaps a table, and a hammock. How little sufficed these children of nature! (Sartorius 1858/1961:6; Müller 1864:203). Several authors recognized how ecologically sensible such a house might be (e.g., Müller 1864:220). However, one could also see through the slits between the poles and that was surely indecent, especially since naked children played among domestic animals behind the rustic lattice-work and *jarochas* went about with little covering above the waist.

Of *jarocho* customs, the travellers noticed mainly the *fandango*. The word seems to have been taken simply as the regional designation for an evening of dancing. We read of the music, the typical steps, the costumes, and the flashing eyes. However, the word has more volatile connotations and embodies, whether the writers knew it or not, some further disparagements. Evidently it was understood in Mexico as something of a scandal. The dancing was lascivious (Deleon 1983:37) and the whole affair would often end as a brawl or a shoot-out (Santamaría 1959:520). In Ferry's fictionalized *fandango*, the dance does end in a machete fight. "What is a fandago without some little quarrel to enliven it?" (Ferry 1856:300).

Several travellers found the goings-on at a *fandango* pathetic—perhaps because they did not stay long enough (Heller 1853:53). The dancing was nothing more than a stamping of the feet. The enthusiasm of the dancers was exaggerated and the music meagre. Yet the dancers seemed quite without care, entirely given over to distraction and enjoyment.

Robertson happened on a country wedding one night during his journey from Veracruz to Jalapa by *litera*. The conveyance was let down for a change of mules at Plan del Río; for half an hour Robertson watched the merry-making in the *venta*, which had been given over to celebration for this evening. He saw "all those connected with the wedding party putting forth a variety of attempts at many-coloured finery." He heard a poorly tuned "jingling guitar" and a singer improvising as best he could to the delight of the celebrants. There was drinking, smoking, "and in a corner cards were calling up the darker passions of the older men." What seemed to impress him most were the attempts he saw to imitate the dancing at an aristocratic ball. "The effect is laughable" (Robertson 1853, 1:272–3).

Müller describes a festivity that he chanced on. He was out hunting with some companions when they heard the sound of a mandolin and the noise of a dance. Through slits in the wall of a house Müller saw what seemed to him an orgy in progress. It was early morning; presumably the celebrants had been there all night. Then he also saw the body of a child laid out amid wilted flowers, its skin beginning to discolour, and realized that this was a wake. The father came out to invite him in, which he declined with thanks, but the mother hurried out to offer him a drink anyway. She was suppressing tears and attempting a smile. What superstitious barbarism! And how Mexican! (Müller 1864, 1:220–2).

EXPLANATIONS

Among the foreigners' observations, especially in the systematic sources, and those of the Mexican observers they cite or that one comes on otherwise, there are attempts to explain some of the characteristics of the rather unimpressive people of the lowlands. Why did they have the attitudes that seemed apparent?

How had the Indians of this region, or the entire country for that matter, reached their low estate? Sartorius had become convinced that they were innately limited, but also that they had been further stultified, indeed made servile, by the set forms of the Christian religion which they did not understand, and by alcohol. He found them suspicious of people with mixed Indian and white background—no doubt because they were themselves liable to cheat. They were distrustful and calculating regarding strangers generally; they tended to speak ambiguously. In these ways they maintained themselves a distinct people within a people.

Heller bested his mentor, Sartorius; at least his observations are more palatable now (1853:57–8). He looked at the Indian physique and found impressive musculature and endurance, as well as decent behaviour and taciturn dignity. Some of the ancient strengths were still detectable, but mainly Heller reflected on how the impressive pre-Contact cultures had been reduced. Their gods had been taken from them, to be replaced by wooden images of saints, but not the knowledge of the true

God. They had been denied education, worked mercilessly, and introduced to rum, hence their pathetic condition. Humboldt had made similar observations in general terms. The Indian was indolent, but tropical nature facilitated this tendency, and centuries of highly unfortunate colonial circumstances had contributed greatly to his extreme misery (Humboldt, Dunn ed. 1972:65).

Various authors came to the defense of Veracruzan lowlanders in general, exonerating the *jarocho*. Villaseñor y Sánchez, the Mexican historian and mathematician who held various high positions in the viceregal administrations of his time, wrote about economic conditions in the lowlands of Southern Veracruz in the mid-eighteenth century (Villaseñor y Sánchez and D.J. Antonio 1746:366). The region was considered potentially very productive, but since there was a limited market locally and none farther afield, it made no sense to work hard or to apply any but the most rudimentary technology.

In 1803 the secretary of the consulado of Veracruz, José María Quiros, again analysed the reasons for agricultural lethargy in the lowlands (Florescano 1976:62–71). The conscription of men from coastal communities around Veracruz for the defence of the entrance to the colony against the English was seen as the principal reason for the low production. Next to it were the high rents and disadvantageous conditions of tenure imposed by the *latifundistas*. Contracts were of short duration and any improvements remained the property of the owner. Humboldt seems to be obliquely citing these very observations, but does not allow them to vitiate his judgment (1811, Book III:256–257).

Richthofen elaborated on rural problems at mid-century, particularly as it effected ranchers (1854:257–60). Cattle were often stolen, one's horses, mules, and fodder could be requisitioned for military purposes. Labour was scarce; those workers one had could be pressed into military service. And then, too, one might well be forced to make a loan for some urgent public purpose.

Koppe, in this regard as in many others one of the most discerning of our foreign observers, remarked on how useless it was for people in the lowland to attempt economic gain in a context of oppression (Koppe 1837, II:169). He also cited a report by Lucas Alamán, rendered when this distinguished historian was minister of the interior in 1830. It outlined the deleterious effects of revolutionary unrest on agriculture and ranching. These included plunder and destruction, of course, but also the lack of labour and the unavailability of agricultural credit. To a greater or lesser extent these retarded all production in the area under consideration throughout the half century after Independence.

Alamán went on to identify what he considered a fundamental problem, intractable during all of the nineteenth century and not really substantially remedied until well into the twentieth: the almost complete lack of roads suitable for wheeled traffic throughout the country beyond just the one road between Veracruz and Mexico City. For all practical purposes the countryside was divided into innumerable unconnected production entities. Under these circumstances a good crop could be a disaster since it would reduce prices severely in the immediate surroundings.

In 1831, José María Iglesias, an official in the Department of Acayucán, southeast of Veracruz, reported general idleness in his region. His observation on the main reason for it is in effect an eloquent restatement of Villaseñor y Sánchez, after a lapse of more than eighty years: "The desperation of the rural people reduces them to indolence. Work has become disgusting. And who would work when perspiration is not repaid"(1966; author's translation).

The desperation went unnoticed by most of our foreign observers. Very few spoke Spanish; most stayed on the main road and crossed the lowlands as quickly as possible. In their accounts we see the material aspects of culture, and only the surface of a very limited range of behaviour. The various reasons given by thoughtful observers for the scanty land use in the lowlands were not widely known or credited. The idea that lax behaviour could be adaptive certainly did not resonate as did the maxim of the lazy native. The accepted wisdom was that lowland resources were very promising, especially when seen from Jalapa, but that the lowlanders and their traditional ways were inadequate for the development of a thriving agricultural economy. Commercial agriculture was needed, but only energetic intervention from abroad or at least from the highlands would bring it about.

Many years after our company of travellers had come and gone, Melgarejo, historian, anthropologist, and Veracruzano, took issue with the way in which the *jarocho* was still often portrayed:

The image of the *jarocho* as a man sleeping in a hammock slung between palms on the fertile banks of a river is an old lithograph, unjust to the fisherman who passes the night in the water, shivering with cold and bitten by mosquitoes, obliged to sleep during the day in order to fish at night. No less false is the calendar picture showing an edenic countryside, rich with crops. It does not take into account the *norte* and the hurricane, the drought severe enough to make spit sizzle on stone, the floods that reach the eaves, the weeds that strangle the crops, provoking insects, noiseless snakes. The temperament of the *jarocho* is irritated by a geography that, although it may be prodigious in material resources and picturesque landscapes, menaces human life day and night with endemic diseases. Distances and abandonment have obligated him to create his own world, to be defensive even though he may have been born with expansive tendencies. He is ridiculed and betrayed. In his anger he takes recourse to his mulatto grammar, and channels his self-defense into sardonic insult or boisterous joking. He discharges his ire in dance and spits blood while singing an old song. (Melgarejo 1979:74–5, translation by the author)

CHAPTER NINE

Above the Oaks

During their ascent out of the humid lowlands and through the "chaparral," stagecoach passengers travelling in the dry season had to endure ten or twelve hours of terrain that was "rugged, sterile, uncultivated and forlorn" (Tudor 1834:178). Then finally, something like two hours out of Jalapa, and about 1,000 metres above sea level, they saw the oaks. The air freshened and everyone knew that the danger of yellow fever had abated. They had entered a region that appeared as luxuriant as the humid lowlands, but benign.

Sartorius provides us with an integrating passage here, a description of the whole sweep of the coastal landscape as he had seen it from the top of some Prehispanic mound, just below the oaks. Looking east he saw immediately before him the sloping hill land, covered as it happened, where he was looking, with "light-green prairie," but which he tells us elsewhere was a mixture of herbaceous vegetation and woody growth, the *chaparral* of the soldiers. This was

succeeded by the dark forests on the coasts. Beyond this the blue gulf is visible, and even the sails of the ships may be discerned; for in a straight line we are not so very far from the sea after all.

If we turn to the west we behold dark wooded mountains, above which, jagged and abrupt, rise the highlands, but to the north and south the mountains extend in beautiful undulations to the distant horizon. . . . Here we can breathe freely, no pestiferous vapours rise from the soil, no intermittent fevers rob the planter of his vigour, no enervating heat hems his activity. A soft mild atmosphere prevails here all the year round, rendered pleasant during the day by the sea-breeze, cooled at night by the refreshing mountain air. Here the clouds driven by the trade-wind towards the highlands most frequently discharge themselves; the country is never long without the fertilizing rain, and the plants are nightly refreshed with a heavy dew. Without artificial irrigation, here flourish the sugar-cane, rice, tobacco, and the banana, without wearisome labour bounteous nature furnishes abundance of wholesome food within a small space. (Sartorius 1961:11–12)

The observations made at these upper margins of the coastal lowlands bring our nineteeth-century rendition of landscape to a kind of culmination. The imagery is given an almost excessive elaboration, the predispositions are restated, revealed more clearly than ever, and with some variations. The ambiguity regarding tropical nature, the dread of poison hidden in the greenery, fades from the accounts. Here now is luxuriance indeed, and one senses from Humboldt and various of those who followed him that such impressions gained above the oaks, particularly with respect to the potential for agricultural production, are projected downslope to enrich the assessment of the lowlands in general. The ethnographic counterpoint persists: even though the people of the favoured region seem more impressive to a number of the observers than the *jarochos* downslope, the challenge of this part of the tropical environment has not been met any more successfully than that of the lands below.

The narrow band between the first oaks and the ascent of the Sierras proper consistently induced euphoria, whether approached from above or below. This was paradise. The concept included luxuriant vegetation and dramatic topography, as well as Jalapa, the city of flowers, the provincial Spanish-American town that was still quite charming but that clearly had seen better days. Actually reading in succession all the descriptions of the region and the town makes paradise tedious. Language obviously failed many of the travellers. There is much repetition and groping for adequate allusions: Eden, Elysium, a bit of heaven fallen to earth, the land of eternal spring and the city of refuge. Madame Calderón de la Barca brings the reader from Dante's purgatory, which is rather an apt designation for all the miles of *chaparral*, to paradise: "The road is difficult, as the approach to Paradise ought to be . . ." (Fisher ed. 1966:67). Robertson, the expansive English traveller, became a little self-conscious after his own effusion and in self-defence offered an anthology of other travellers' descriptions (1853, 1:291–9). An American officer formulated the essentials: "Jalapa is deservedly esteemed one of the most pleasant towns in Mexico, and a more delightful climate and more beautiful scenery can scarcely be imagined. It is situated on a side hill, and is surrounded by fine groves and well cultivated fields. It contains about ten thousand inhabitants, and is celebrated for its handsome women and flowers, which here reach the highest perfection" (Brackett 1854:64).

The town and its setting elicit ritualistic adulation from natives and visitors to this day. On a clear morning, away from the clamour of the throttled, sprawling city, say somewhere among the coffee planatations just to the south, adulation still comes easily. No one has outdone Jalapa's own Leonardo Pasquel, who for many years published Veracruzan lore from his Editorial Citlaltepetl, the name itself a tribute to Mount Orizaba, in Nahuatl. Euphoria surfaces in more serious contexts too. At an anthropological conference held in Jalapa in 1952 on the subject of "Huastecos, Totonacos y sus Vecinos," one of the participating scholars offered an introductory essay in praise of Jalapa. After a witty review of the efforts of some of the nineteenth-century travellers to do justice to the town, he made his own attempt, ending with the affirmation that for a few days the participants had an opportunity, rarely granted to mortals; they would be living in Arcadia (Martinez del Río 1952–53:25). He was followed by a paper on the "Anthropogeography of Central Veracruz" by William T. Sanders (1953), which must have brought the discussion down to earth. His hard-hitting, unromantic, and "ecological" approach to Mesoamerica was already very apparent.

THE NATURE OF PARADISE

The modern traveller going upslope in the dry season can still easily understand how his nineteenth-century colleagues felt. Cooler, less oppressive air comes through one's windows at an altitude of about 1,000 metres. This is in the vicinity of the old inn of Lencero, the remains of which are within what is now called Dos Ríos (Figure 16a). There is no line on modern climatic maps that quite coincides with this sensation, but average annual and dry season temperatures do decrease notably upslope. An increase in total and dry season precipitation is apparent as well. Southeasterly trade winds bring wet season (June to October) rains here as elsewhere in the lowlands. Cooling and hence precipitation are facilitated by the air mass's movement upslope. Also the mountains blocking the moisture-bearing *nortes* from the terrain to the east are so aligned that they do not have this impending effect above the oaks, thus allowing some precipitation throughout the dry season as well. Garcia's regionalization of mid-slope climate, the best available, is necessarily a web of transitions ("semicálido," "subhúmedo"). To the travellers it was all much more distinct.

The travellers' usual associations with the easing of the atmosphere,

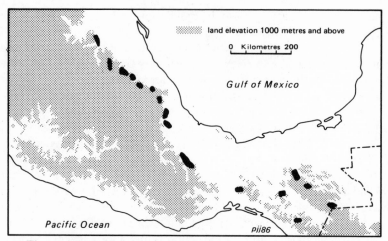

16a The environs of Jalapa—a map of paradise

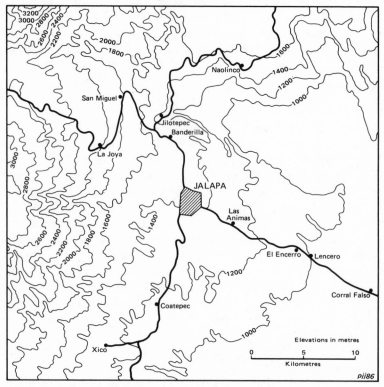

16b Regions in which temperate and tropical vegetation mingle, after
Miranda and Sharp (1950:315)

both on entry and exit from the environs of Jalapa, were found in the region's plants, which are transitional as well, in several senses. A recent authoritative designation is very general and intimates ambiguities: *bosque mesofilo de montana*, that is, simply mid-slope forest, growing under humid conditions, but not so humid as those of the lowland rain forest. This vegetation has been given other names, such as cloud forest, mountain rain forest and mixed deciduous forest (Rzedowski 1978:156–7; Zolá Báez 1980:34).

The environs of Jalapa, as indeed a whole series of habitats in roughly the same altitudinal band along the eastern side of the Sierra Madre Oriental, happen to be a meeting place of genera with varying geographical affinities (Miranda and Sharp 1950). A good number of these are related to vegetation of extra-tropical regions to the north. In fact, these give the plant community its dominant characteristics. They include the two signature genera, *Liquidambar* and *Quercus*. The first is notable for its tall, full, conical form, similar to various deciduous trees the travellers were acquainted with in their homelands. The second, of course, was familiar too, and a particularly welcome sight. Genera with tropical affinities are more numerous, but less obtrusive. They often make up the lower stories, reflecting micro-climatic variations: mainly somewhat higher and more even average temperatures in the shade.

The entire picture presents an infinity of variations. The lists of species are numbing, and still not completely known. Some are evergreen, other deciduous. However, this is no simple contrast; the second condition, at least, is a matter of degree. Furthermore, this region has long been subject to human disturbance. A great number of succession species and cultigens—some temperate, some tropical—add their complications. Coffee is at present the most notable crop; it has its altitudinal limits within the zone and is sometimes used to designate it. Travellers often include wry asides about the most notable commercial plant of their time: *Ipomoea purga*, sometimes called the jalap root, a trusted friend of old-school physicians. Schiede, the indefatigable botanist, searched the plant out and found it grew wild in the forest just upslope from Jalapa. Its tubers were dried over domestic cooking fires and then brought to Veracruz for export. He expected that someday it would be grown commercially on a large scale in temperate lands, which has not happened (Schiede 1830:473–4). The purgative is evidently still in use, but the root has long since become difficult to find around Jalapa (Zolá Báez 1980:2).

The environs of Jalapa are transitional in another sense: between a pine-oak forest upslope, and the tropical deciduous forest downslope, or *selva baja caducifolio*, which the soldiers called *chaparral*. Looking more closely, it becomes apparent that there are transitions to transitions

(Zolá Báez 1980:Figure 5; Gómez-Pompa 1978), one of which was critical to the perceptions of the travellers. The tropical deciduous forest merges upward into oak forest (*encinar*), and that in turn into mesophytic forest proper, also called "mixed deciduous" or "cloud forest" or "montaña." The *encinar* is made up of various species, dominated by *Quercus peduncularis*. These trees are not very tall—from eight to fifteen metres—but they grow in scattered stands, are rounded in form, and evergreen (Zolá Báez 1980:60). They are therefore easily recognizable as separate entitites by the passing traveller. The lower limit of the oaks may be detected on air photos as a crenellated line, extending hundreds of fingers downslope along shallow dips between dryer, scrub-covered hills. They seem to reflect a downward drift of cooler, moister air along the valley bottoms.

The oaks were usually noticed in the vicinity of the inn or *venta* of Lencero. The modern road divides the courtyard and lodgings from the building in which the mules were stabled, as the old road did, but the traffic hardly slows down now. This was one of the earliest post-Contact inns along the road. It is named after a soldier in the army of Cortés who took in washing and was nicknamed Lienzero, after *lienzos* or clothes (Melgarejo, personal communication 1979; Bermudez 1977: 206–7). Nearby are the buildings that remain of Santa Anna's hacienda, El Encerro, a corruption of the name of the *venta*. There are a few oaks still in the hollows, but since this wood makes good charcoal, they have been much depleted (Figure 17).

To the travellers of the nineteenth century the first oaks of the *encinar* were one of the sharpest environmental markers in the lowlands. They signalled a relief from the dread of yellow fever—a partial relief, because one could still succumb to the effects of an infection sustained downslope. Only after several days could one be sure.

The opposite, western entrance to paradise also consisted of an easement and change in the vegetation. These were usually sensed about fifteen kilometres northwest of Jalapa, just below the village of San Miguel, now just off on the north side of the straightened new road. The gradient was particularly steep here and the American drivers brought the stagecoaches down it at a frightening speed. The fogs that frequently surrounded the pass often dissipated here as one descended, the chilly mountain air softened, the evergreen forest of oaks and conifers gave way to a mixture of deciduous species; soon one saw fruit trees and flowers.

PERCEPTION OF THE FAVOURED REGION

The travellers' effusions may be reduced to several essentials. One of these was an amazement at the variety in the natural vegetation, and

17 The oaks that signalled safety from infection by yellow fever

Sartorius said it best. He had found the plants of the humid lowlands impressive, but here around Jalapa there was real luxuriance: "all life and organic activity." Trees, palms, grasses, agaves, countless lianas, and ferns were noted, especially the ferns: "Fancy . . . such groups of slender stems, twenty or thirty feet high, surmounted by a gigantic fan of fine pennated foliage, spread out like a vast parasol. How beautiful is the blue sky through the delicate texture!" (1961:12). Lichens and mosses clothed the rocks, epiphytes of many kinds swarmed over the trees. Grasses and sedges and other ground cover filled the clearings. Wild fruit was freely available in the forest. And so on—all perfectly plausible enthusiasm from someone especially interested in plants.

Schiede went out from Jalapa in three or four directions to observe and to gather specimens (1828, 1829); various other travellers also went on excursions from here, but mainly for recreational reasons. Schiede was single-minded, a naturalist completely absorbed, but weaving the species into a flowing prose description, when others might have made lists. He records only the briefest impressions of settlements and views, mentioning the inhabitants only as they helped him get what he wanted. A sensitivity to altitudinal variation, to the transitions within transitions along these mid-mountain slopes, permeates the descriptions of plants. In one of his few ethnographic observations he tells of Indians north of Jalapa who live in the cooler heights among the cloud forest, but descend along difficult trails into the head of a valley in order to raise bananas lower along the profile. The forest thrilled him in its entire assemblage. He could imagine nothing more luxuriant. It was so different from the coniferous forest of temperate latitudes, with its solid stands of single species, the trunks mostly clean of parasites and the floor mostly clear of underbrush. Here, wherever one looked there was life. He went with a will at the detailing of this life, but then looked up, as it were, in one affecting passage and despaired of doing justice to the whole (1828:235).

A modern visitor needs to go further afield, but can still gain similar sensations around Jalapa. The mid-slope band is still an excellent region in which to botanize; witness the botanical garden and research facilities of the Instituto Nacional de Investigaciones sobre Recursos Bióticos just south of Jalapa.

The travellers found great variety not only in the natural vegetation but in the cultivated plants as well and were especially struck with the mingling of useful plants out of both tropical and temperate lands. This was obviously the best of all possible worlds. In the fields surrounding Jalapa, maize grew next to wheat, although the latter seldom ripened. In the gardens coffee and sugar cane ripened next to cabbages; melons and artichokes were juxtaposed with common pumpkins and onions; apples and peaches grew next to oranges, limes, grapefruit, and bananas; cacti and agaves interspersed with flowers known in temperate homelands (Mühlenpfordt 1844, II:83).

Often the flowers of this exhilarating land had to be given their own special burst of prose: "Large scarlet blossoms, and hanging purple and white flowers, and trees covered with fragrant bell-shaped flowers like lilies, which the people here call the *floripondio*, together with loads of double pink roses that made the air fragrant as we passed; and here and there a church, an old ruined convent, or more rarely a white hacienda or farmhouse" (Calderón de la Barca, Fisher ed. 1966:73).

Implicitly or explicitly the travellers indicated relief at the disappearance of the malevolent aspects of the tropics. There was marvellous variety in nature all around, but not, it seemed, the same preying of life on life that had been noted in the humid lowlands; a fertile, well-watered soil, but no evil vapours. These, of course, were some of the most attractive aspects of both the garden out of which Adam and Eve had been driven and the paradise that awaited the faithful.

The fact that Jalapa and its surroundings often have most disagreeable weather could not much alter the paradisiacal impression. Fog,

light rain, and decidedly chilly temperatures accompany the dry season
nortes; fog and heavier rain are also brought in at intervals by the trade
winds during the rest of the year. Some travellers did pay their tributes
to such weather, but it did not dampen their enthusiasm very much: "It
was raining, and continued to rain all day, not a slow, dreary drizzle, nor
a torrent of heavy drops, as rain comes to us, but a gauzy veil of moisture that scarcely stirred the grass on which it fell or shook the golden
pollen from the orange-flowers. . . . We could not stroll among the
gardens or sit under the urns of the Alameda, but the towers and
balconies were left us; the landscape, though faint and blurred by the
filmy rain, was nearly as beautiful, and the perfume could not be washed
out of the air" (Taylor 1850,II:326).

Payno tells how he was huddled in his lodging, with no desire to go
out. The sky was the colour of lead, clouds rolled over the roofs, the air
was dense with fine droplets, and the stones of the uneven streets trea-
cherously slippery (in Tavera Alfaro 1964:86). At such times, Jalapeños
muffled themselves in their serapes and muttered, "Ave María purísima,
que venga el sol" (Robertson 1853, I:298). Nevertheless, the town and its
surroundings were known to be balmy and colourful, even if they were
found to be cold, grey, and miserably wet. And usually the skies did
clear at some time during any given visit, as they did for Payno, confirm-
ing what he and most other educated Mexicans knew about this part of
their country.

It was remarkable that the people improved with the quality of the air.
In one of the superficial accounts included in this study as "controls,"
the point is made baldly: "The inhabitants of the district appear hand-
somer, better dressed, and more intelligent" (Mason 1852:9). Several of
the more astute observers said similar things but in subtler ways: "As we
approached Jalapa, the road was thronged with people hastening to
church, it being Sunday. We were delighted to see them so well dressed
and cheerful, and we were cheerful ourselves because we had reached
the region of health" (Gardiner 1959:26). Blanchard described what
was, essentially, a picturesque European rural landscape, complete with
order-loving people (Blanchard 1839:113). For Taylor "the whole landscape was like a garden . . . the fields of the freshest green, the groves
white with blossoms, and ringing with the songs of birds, and the
gardens loading the air with delicious perfume. Stately haciendas were
perched on the vernal slopes, and in the fields; on the roads and in the
winding mule paths of the hills we saw everywhere a gay and light-
hearted people . . . the time went by like a single pulsation of delight"
(Taylor 1850:325).

Nevertheless, as though on reflection, some travellers still found it
necessary to indict the inhabitants for an insufficient use of the

resources of their favoured habitat. Sartorius took Humboldt's dictum on the debilitating effects of an over-bountiful tropical nature to apply particularly to this mid-mountain zone. Its inhabitants enjoyed what was available and gave too little care for the future. The birds confirmed them in their thoughtlessness; nests were constructed less artistically than in colder regions. A cultivated field in this context appeared like an oasis in a wilderness (Sartorius 1961:13–14).

Becher put the indictment in another way and drew some conclusions. Like Humboldt he viewed the surroundings of Jalapa from the top of the venerable monastery of San Francisco (modern travellers may find a very similar view from the terrace of the Hotel María Victoria's restaurant). It was a fine panorama, but not yet fully animated by man, a landscape rich in its natural vegetation, but relatively inert. Only with a liberal political system, religious freedom, and the resulting immigration would this enchanting Eden come to its own (Becher 1834:47).

The American traveller, Wilson, was amazed at the vast areas of land just east of Jalapa, "with a soil the richest in the world" (1855:49), given over to grazing. The implication was that surely something better could be done with it. That thought had already occurred to a fellow countryman of his, the American diplomat Waddy Thompson. The year was 1842 and the immediate objectives of "Manifest Destiny" were Texas and California, but the rest of Mexico was fair game too in the minds of the venturesome. He was moved to predict that "No spot of the earth will be more desirable than this for a residence whenever it is in the possession of our race, with the government and laws which they carry with them wherever they go. The march of time is not more certain than that this will be, and probably at no distant day" (1846:13).

THE GEOGRAPHY OF EUPHORIA

Taylor tells of an evening's walk by a stage load of travellers unavoidably delayed for a few days in Jalapa on their journey downslope. They went out on the day of their arrival to promenade with the local people on the outskirts of the town. They became quite euphoric and immensely talkative—in four languages. One of the travelling company was moved to declare that his pulse was quicker and his blood warmer than it had been in twenty years: "We talked thus till the stars came out and the perfumed air was cool with invisible dew" (1850:326).

Such reactions are recorded repeatedly in the accounts and seem to have a geography, a locational rationale. Many of the travellers coming via Jalapa traversed other mountain slopes in their itinerary through Mexico and noted similar environmental conditions and reactions elsewhere. Altitudinal zonation was described along the old alternative route through Córdoba and Orizaba and up the escarpment, but sensations do not appear to have been as sharply defined. The road does not cross vegetation boundaries as perpendicularly and topography complicates the picture. Orizaba received its praise then and still does, but it has not been able to rival Jalapa in environmental attractions. Commerce and industry, of course, have been another matter.

Beltrami noticed west of Tampico and just north of the present Ciudad Mante how, on ascent from a vegetation very much like the *chaparral* of the American soldiers into an environment like that of Jalapa, tropical nature was reanimated, and disenvenomed. Luxurious leaves replaced bare branches and thorns; snakes were fewer and birds more numerous; the streams had fine fish in them and not crocodiles; a singed landscape had given way to freshness (in Glantz 1964:183–4). Tayloe felt a reanimation coming downslope from Zacualtipán, an important colonial way station betwen Pachucá and Tampico: "The temperature is delightful. . . . Eggs and bananas are abundant—the water is excellent, and the people, very few individuals of them speak any Spanish, are very obliging, and attentive to our wants" (in Gardiner 1959:183).

A cartographic summation of the favoured, narrow mid-slope band in the Sierra Madre Oriental would be difficult. Neither contour intervals nor isohyets nor isotherms serve very well. The closest approximation is perhaps Miranda and Sharp's map of "certain temperate regions of eastern Mexico," the regions where they found the remarkable confluence of temperate and tropical plant genera (1950:315; Figure 16b).

Several travellers took the geography of paradise farther afield. Madame Calderón de la Barca made a comparison between the environs of Jalapa and various locations in Michoacán (Fisher ed. 1966:609). Bullock described the striking effect of going out of the environs of Cuernavaca down to Acapulco, a more abrupt zonation, he felt, than along the Jalapa road (1866:22–3). One could trace a similar zonation along the western and eastern slopes of the Andes. Wherever one comes upon such favoured regions, one is likely to find certain historical geographical commonalities. They offered choice settlement sites for the early European immigrants. Sugar and coffee became their important export crops. For the affluent living in cities below, they have long provided a refuge from heat and disease. For those living above, they offer relief from cold and meagre oxygen. Modern urbanites find in them excellent locations for weekend or summer homes. These regions do not, however, attract massive tourism as do the tropical beaches and the monumental ruin sites. They are not easily "sold," it seems, except to the sophisticated traveller.

JALAPA

The travellers created a town to conform to their expectations and the demonstrable, bracing effect of this environment. They were more relaxed than they had been in Veracruz, but most had even less time in this town; it was usually just a stopover on the coach trip into the interior. If one remembers as well that very few spoke Spanish, it is very clear that contacts were limited and that there was virtually no opportunity for incisive inquiry. The result was charming, with only the odd intimation of difficult living conditions under the tiled roofs and just behind the white walls draped in bougainvillea. A few detected decadence among the well-to-do, and with varying degrees of explicitness used that to elaborate on the indictment not so much of the lowlanders but Mexicans in general.

Jalapa had seen better days. Various Mexican historians have described what Jalapa was like during its golden years (1720–78), the epoch of its famous fair (Carrera Stampa 1953; Trens 1955). The fair provided a context, above the range of yellow fever, for the distribution of the goods brought by the periodic fleets from Cádiz. Its site had been a subject of the ongoing competition between Orizaba and Jalapa, which had already involved the routing of early roads and would include the routing of the new road over which Humboldt was to come as well as the railway that would alter the pattern of movement through the lowlands in the second half of the nineteenth century.

The fleets arrived at irregular intervals. Advance notice came by letter from Spain; the actual arrival at the anchorage off Veracruz was signalled all the way to the capital by fires. Soon traders from the interior, crews of the incoming ships, muletrains from both directions, and all sorts of outsiders who intended somehow to make a profit, converged on Jalapa. The town was inundated; for two or three months all was clamour and enthusiasm over the fifteen to thirty million pesos that were changing hands. Immediate gain accrued to Jalapa through the expenditures on the storage of goods and mule transport. The trading was attended by a great deal of religious ceremony, for which the merchants rewarded the local clergy handsomely. As the eighteenth century progressed, the townscape filled out with new *almacenes*, private residences of wealthy merchants, as well as new rental housing and store space. Many farmers in the surroundings turned from activities such as the growing of tobacco, to commerce; speculation in real estate was rife. The tastes of the townspeople became more sophisticated with the handling of European goods and ideas.

With the beginning of the liberalization of trade the need for one central fair disappeared. An economic recession followed quickly in

Jalapa. Many trading houses transferred their operations to Veracruz or the capital. The unrest attendant on the struggles for independence reduced traffic along the road and made matters worse. By the early nineteenth century Jalapa's function as a supplier of mules and *arrieros* had severely diminished. It was not helped much by the subsequent establishment of carting and stagecoach lines, and certainly not by the railway that eventually diverted traffic along the Orizaba route.

Jalapa's shops and warehouses were not impressive when Bullock explored them in 1823 (1825, 1:50). He noticed imported manufactured articles of poor quality being sold for three and four hundred per cent of cost. Jalapa did, however, have some fine barber shops and extraordinary laundry facilities. Evidently some people from Veracruz sent their laundry up to be washed here. The town's open market was repeatedly a source of amazement. Tropical and temperate produce was available in profusion, as were drinks cooled in urns sunk into piles of wet sand and, indeed, the whole range of common Mexican cooked food. Mayer, who left one of the good descriptions of the Jalapa market, was led to the undoubtedly exaggerated conclusion that "with the great mass of Mexicans there is no such thing as domestic cookery" (1844:16). For a few coins, they could get everything they needed at a succession of foodstalls.

A correspondent from the *Boston Advertiser* described the Sunday market about two weeks after the battle of Cerro Gordo (Smith and Judah 1968:220-1). The country and the region were prostrate, but the plaza opposite the cathedral was so filled with the vendors of foodstuffs that one had only just enough room to step between the heaped stocks in trade. The correspondent, of course, spoke as one who had been on field rations for weeks. Here spread before him was fresh food of every kind, which he proceeded to itemize with relish. There were flowers and fuel for sale as well. In the shops surrounding the plaza he saw stacks of staples: rice, sugar, corn; and he even found fresh cow's milk.

No doubt the richness of the Jalapa market did reflect the productivity of its immediate environs, especially the farmland to the south. Local trade seemed to persist at least intermittently in spite of generalized national disorder. However, the market was probably often perceived as rich because it was an exotic spectacle, and thus misleading if taken as an indication of general economic health.

There are many indications in Rivera Cambas' rambling history of Jalapa in the nineteenth century that in fact the living conditions for the majority of the population were often desperate. Bare necessities were frequently scarce, prices and unemployment high, as a result of the interruption of movement along the road. Municipal coffers were chronically depleted and political conditions often approached anarchy. However, all this seems to have stayed largely beyond the ken of the visitors. In any case it was not allowed to dull the imagery. There were various Jalapas, then, as there had been various Veracruzes.

Mrs. H.G. Ward, wife of the English chargé d'affaires, graced the second volume of her husband's Mexican memoirs with a drawing of Jalapa as seen from a high point southeast of the centre (Ward 1828; Figure 18). It is expertly drawn, rich in nuance, the best of a series of nineteenth-century representations of the town, and easily the equivalent of the best attempts to describe the scene. It illustrated an authoritative account and hence became widely known. Natural spotlighting was allowed to set the town off against its surroundings. This is theatre but also a frequent actual occurrence in an area that has patchy cloud cover on many days throughout the year. Varied vegetation surrounds the town. The whole scene is an idyll, another evocation of the *picturesque*.

It has been noted how the townscape of Veracruz had stimulated comparisons with Alexandria and how the dunes behind the town had seemed like an Arabian desert. The eastern metaphor was continued in some of the descriptions of Jalapa. Blanchard found Moorish elements in its principal buildings, together with Romanesque and Gothic (Blanchard 1839:106). Altamirano extravagantly transformed the town into a graciously reclining oriental maiden, emeralds setting off her white clothing and flowers at her breast—a worthy match for the drawing by Mrs. Ward (in Tavera Alfaro 1964:339).

The drawing has a more prosaic companion-piece in a map of the town as it was in 1776, the year of the last of the great fairs and more or less as it will have appeared to the nineteenth-century visitors (González de Cossio 1958). The map is dedicated to the consulado of Cádiz, which figured importantly in the management of the fairs, and seems to have been presented to them as a souvenir, with presentiment of the liberation of trade in 1778. It is richly detailed, including the location of the Spanish merchants' houses and a notation at the bottom of the legend that this last fair had yielded twenty million pesos. It illustrates the usual concentric zonation of the Spanish-American town, from the monumental buildings of the centre, the houses of the principal families, through lesser buildings, still of masonry and tile, to the wood and thatch houses on the outskirts, most of the latter pertaining to Indian *barrios*.

Tayloe, on approaching, found it difficult to understand how Jalapa could accommodate the population that it was reputed to have, and his editor adds as a footnote that "in all ages, past and present, the non-Latin traveller has found it difficult to comprehend the density of population in Mexican communities" (in Gardiner 1959:29). Citing some

18 View of Jalapa, by Mrs. H.G. Ward, in Ward (1828)

actual population figures for Jalapa has to be the same gingerly exercise
it was for Veracruz. The resident population of the town at the turn of
the century was estimated by Humboldt to be 13,000 (Humboldt, 1811,
III:269). Rivera Cambas reported a population of 9,523 for 1820 and
10,428 for 1826 from sources known only to him, but judging by the
illusory specificity of the last digits, almost certainly from some official
report (1973, v:19). For 1831, Koppe cited a total population of 10,628
from a governmental source (1837, I:157). Wappäus concluded in 1863
that the town had somewhere between 1,000 and 15,000 people,
depending on whom one reads (1863:154). These figures suggest a post-
Independence decline, similar to but not quite as drastic as that of
Veracruz, and a slow subsequent growth, again as in Veracruz. Mrs.
Ward's Jalapa seems far too small for even the lowest of these numbers.
However, this is not to be attributed only to the structure of the Spanish-
American towns. A great deal of the townscape was left off, to the right,
as the map of 1776 shows.

Several authors refer to fluctuations in the population. In 1825, when
Veracruz was under the threat of bombardment from the Spaniards still
occupying the fortress of San Juan de Ulúa, Tayloe had found it almost
deserted and the population of Jalapa correspondingly augmented (in
Gardiner 1959:29). To a war correspondent from the *Boston Advertiser* it
was clear that the population of Jalapa had almost doubled since the
Americans had threatened Veracruz. This was undoubtedly exaggerated, but the immigration was sizeable and it pointed up Jalapa's old
function as a refuge. Rivera Cambas confirms the massive exit from
Veracruz and then recounts how Jalapeños had attempted to raise some
money to aid the refugees. There was almost nothing to be had from
religious, governmental, or private sources; the entire economic context
was desperate (1973, IX:189–92). Each year when the *nortes* ceased and
yellow fever began to threaten, those who could afford it migrated to
higher ground. Foreign merchants with *almacenes* in Veracruz maintained residences in Jalapa.

Of the two major landmarks of the townscape Mrs. Ward recorded,
only one remains. The prominent spire on the right belongs to the
Parroquia de San José, which is still Jalapa's major church. The Con-
vento de San Francisco, the massive complex to the left, was founded in
the middle of the sixteenth century, its central structure intended to
serve as a fortress as well as a church. Humboldt described the magnificent view he obtained from the top of this building (Humboldt, 1811,
III:268). By the 1880s it had fallen into a ruinous state and was demol-
ished.

Veracruz had seemed monotonously regular, its streets on a rigid
grid, its houses flat-roofed, mostly of two stories. Jalapa was repeatedly

compared favourably. It had an essentially quadrate plan as well, but considerable local relief and several somewhat irregular key streets made it seem, then as now, to be a town of crooked streets and infinitely varied perspectives. Most of its houses were single-storeyed, gabled, and roofed with tiles; walls were white and eaves projected over sidewalks, offering shelter from the sun and the rain. Upland towns of today that are not yet as extensively changed from their colonial form as Jalapa, San Cristóbal de las Casas in Chiapas, for example, allow one to visualize the Jalapa through which the nineteenth-century travellers came.

Latin American towns and cities with substantial remnants of colonial architecture offer the visitor now, as they did then, a virtually interchangeable repertoire of urban forms and indeed "sell" this resource in all sorts of ways (e.g., Figure 19). Spires, facades, cupolas, statuary, heavy white walls, tiled roofs, cobbled courtyards, fountains, balconies, huge wooden doors, and iron grillwork are its widely familiar components. Ferry explored all this on foot in Jalapa and found "at every step some charm" (1856:277). The mountains, the trees and flowers in the gardens, and the architectural detail combined in some new way around every corner.

A number of the travellers found the windows of special interest. They were barred, of course, which made the houses look to some like prisons (e.g., Tayloe in Gardiner 1959:28). However, this was not usually the point. Behind the bars were the invariably beautiful women of Jalapa, one of the sights of this journey, like the peak of Orizaba from offshore, the Puente Nacional, or the first oaks. They had some flaws: they smoked, for one. Bullock suggested they were liable to gossip and a little too demonstrative with each other (1971:53-6). Madame Calderón de la Barca was disgusted with the fact that the women of Mexican high society generally were incredibly indolent, and that they never read (Woolcott 1925:129). The prevailing view was otherwise: "The stately step, the liquid eye, the pale yet brilliant cheek, and an indescribable look of tenderness, complete a picture of beauty rarely matched in northern climes and elsewhere unequalled in Mexico" (Mayer 1844:16).

It was deduced that a humid atmosphere probably helped to produce that delicate complexion (Mühlenpfordt 1844:11, 82). And how could people living in a land of mild temperatures and roses be severe and taciturn? (Payno in Tavera Alfaro 1964:91).

At night the windows became showcases. Ferry noted how

every open window then sends forth a welcome ray of light into the dark and silent street, and the traveller can not but hear the joyous merriment that is going on within. In the warm nights of this beautiful climate the stranger can thus share in these fetes every evening;

19 An excerpt from the repertoire of Spanish-American architectural forms, re-worked from a menu of the Hotel Salmones in Jalapa

he can see the Jalapeñas display their charming vivacity without affectations, from the first moment that the fête commences till the flowers in their head-dresses wither, the harp ceases to be heard, and the windows are closed behind their iron bars. (1856:278)

Protestant visitors walking the streets of Jalapa were repeatedly stimulated to recall their faith's indictment of Roman Catholicism (Ortega y Medina 1955:95–125). This had been begun in Veracruz and would be continued along the journey into Central Mexico, but it was in Jalapa, with its easier atmosphere and invitations on every hand for leisurely, observational strolls, that the indictment seems to have been given some particularly strong expression. Soon after entry through the eastern gate one traversed the Street of the Precious Blood of Christ; this was sacrilege (Tayloe in Gardiner 1959:22). The monastery of San Francisco, which dominated the townscape and offered from its roof a prime vantage point, induced recall of what was perhaps the most scathing representation known in the English-speaking world of Catholicism in Spanish America, the seventeenth-century travel account of Thomas Gage. Some travellers merely referred to this venerable source; the American lawyer, Robert A. Wilson, could not resist copying long excerpts from it into his account. Gage had expected austerity in the monasteries of the New World. He found excellent food, and a prior who preferred sweets and a guitar to books of theology. Monks who had vowed to wear rough wool next to their skin and never to ride, sported silk shirts and went about on excellent mules. During an evening of gambling at the monastery in Jalapa a monk sarcastically referred to his vow never to touch money and scooped up his winnings with his sleeve (Wilson 1855:64–7).

Should a Protestant traveller happen to be on the street when the Host, itself a doctrinal affront, was being carried by in procession, he would be prudent to kneel down on the pavement together with everyone else, for to remain standing might very well provoke a stoning (Tayloe in Gardiner 1959:28, 30–1). The same Mexicans who knelt piously under these circumstances thronged the streets of a Sunday evening in search of entertainment—surely a profanation of the Lord's day.

Waddy Thompson, never hesitant about a strong statement of opinion, set down the basic Protestant position: "To one educated in the unostentatious purity and simplicity of the Protestant religion, there is something very striking in the pomp and pageantry of the Catholic ritual as it exists in Mexico, and I must say something equally revolting in its disgusting mummeries and impostures, which degrade the Christian religion into an absurd, ridiculous, and venal superstition" (1846:vi).

Moreover, many aspects of the Mexican attitude toward Protestantism rankled deeply. The right of Christian burial, for example, had been conceded to them in Mexico only after difficult negotiations. Nothing could induce the national government to allow the building of a Protestant church. All this was obviously bigotry of the worst kind.

The people of Jalapa were observed in passing, very much as the people of Veracruz or the inhabitants of wayside villages and ranchos.

The travellers described society and culture as apparent to them in the market, the crowded evening streets, or at the public events that happened to coincide with their visit. Those lodged in the inns had little opportunity to find out what went on in the homes beyond what they might glimpse through a window or an open doorway. Dignitaries and those well connected were asked to stay in homes, to participate in fêtes, and go out on excursions. During the American invasion, the higher officers took up quarters in homes, while the other men camped out. What these privileged visitors recorded is mainly life among the well-to-do, that is, among the creoles and foreign residents of substance. In total, it is not the *canaille* who are obtrusive, as they were in the descriptions of Veracruz, but the better people.

The correspondent from the *Boston Advertiser* described the coming and going on the central streets of Jalapa a few days after the defeat of Cerro Gordo:

> You pass gentlemen in large broadcloth cloaks, thrown over the shoulder a l'Espagnol—now and then a Mexican officer, mingled with tradesmen and country people in short jackets of blankets; women in coarse mantles, with baskets of produce on their heads; boys selling cakes and candy, and the only thing which reminds you of being in an ememy's country is meeting here and there a soldier, or crowds of slovenly looking volunteers, or passing a sentry in his pipe-clayed belts, quietly pacing in front of the quarters, his burnished musket glancing in the sun, or ringing as he salutes a passing officer. The streets are often crowded with large wagons, conveying the subsistence and stores of the army; little Mexican horses, gaily caparisoned, the saddles often mounted with silver; droves of pack mules, in strings of five or six, the halter of each tied to the braided tail of his "illustrious predecessor," and donkeys almost entirely concealed beneath immense bundles of straw or forage. . . .
>
> But it is on Sunday that the plaza should be seen, and the view is then most animated. This is the principal market day, and the whole place is covered with the vendors of comestibles, seated flat on the pavement, each by his or her little stock, which they bring on their backs from the country. These people have strongly marked Indian features, and dark complexions; the men dress in jackets or blankets, wide trowsers, and large straw hats; the women in a light upper dress of cotton (camisa) with or without a coarse "rebosa" or shawl, and skirts usually of brilliant colors. . . .
>
> On Sundays too one sees what is a rare sight on other days—the ladies of Jalapa, picking their way across the market place to the

church. Many are of unmixed Castilian descent, and quite beautiful. . . . The bells are ringing with redoubled energy, and there goes
the padre,—the corpulent gentleman in the blue gown and broad
brimmed white hat. (Smith and Judah 1968:220-1)

Some of the most interesting observations on society and culture in
Jalapa come from William Bullock, the traveller who professed no business nor diplomatic interests, only curiosity, and who also came with
some effective letters of reference. The people of substance, he notes,
"are patterns of politeness, full of compliments, and profess that their
houses are at your service, but seldom ask you in" (1971:52). Such
conventions bothered some other travellers, too; this was obviously
something the outsider must not be taken in by. When Bullock did get
opportunities to converse (he must have spoken Spanish reasonably
well or had the help of an interpreter), he found that both men and
women were very ill-informed about European affairs and had entirely
exaggerated views on the political and economic importance of Spain.
This would change, he hoped, and the darkness under which Spain had
kept her overseas subjects would surely soon be lifted. He was honest
enough to admit, however, that there was great ignorance in Europe
about Spanish America. Few in Great Britain had ever heard of names
such as Puebla or Guatemala.

Bullock tells about the stir he created with some of the items he
carried in his luggage: a walking stick gun, a portable table and chair, a
camera lucida, "and other little specimens of English ingenuity"
(1971:54). His hosts were astonished at prints of public buildings in
London that he had with him. He heard them exclaim among themselves, "and yet these people are not Christians" (1971:54). The ladies
were devastated by a volume of fashion plates. When he came downslope again in just under six months, to embark for home, he was struck
with a remarkable alteration in Jalapa. Ladies who had formerly
dressed pretty universally in black were now promenading "in the last
fashions of England, in white muslins, printed calicos, and the manufactures of Manchester and Glasgow" (1971:484). The impressions
gained from his volume of fashions, and the wardrobe of an English
lady who had arrived in Jalapa in the meantime, handed quickly
around for patterns, had worked the change. Here, obviously, was a
promising market for British goods.

Even though time was short in Jalapa for most of our observers some
of them were invited to social events among the well-to-do. In the
descriptions of what happened various shadings are added to the
euphoric imagery. It had already been qualified: nature around Jalapa

had been pronounced Edenic but paradise would only be fully realized when there were more cultivated fields, more economic activity, more bustle. It was further qualified now by decadence.

Poinsett, the American ambassador, was taken to a party in Jalapa, where he soon noticed a remarkable aspect of Mexican socializing that persists to this day: the segregation of the sexes (1824:27–28). He somehow found himself first among the ladies, who were playing guitars and singing quite agreeably, but was soon ushered through the patio, up a narrow passage into a cavern-like room where the men were deeply into a game of *monte*. Poinsett, who maintained he never touched cards, was appalled at the way money was being won and lost. Others were to make that same point repeatedly. Here was one major reason for the unfortunate state of the country.

Even more appalling were the animal sports that various visitors were invited to watch. Bullock was treated to some bull-baiting and a gory bull fight, which he was pleased to pronounce very un-English (1971:59–60, 461–5). Kenly and his fellow officers watched something like the calf-throwing seen in North American rodeos as well as bull-fighting, complete with the disembowelment of horses. When an American circus on its way from Veracruz to the capital put on a performance at Jalapa, the officers and men were delighted to be treated to this less barbarous entertainment from home (1873:379–83).

Sarcasm must have been difficult to avoid when confronted with affectations of high culture in a country newly opened to freer influence from abroad. One visitor was not successful. Becher attended a performance of *Othello* in Jalapa and another of something called a ballet. Both were terrible. In Jalapa one should not expect to find one's entertainment in the theatre, but in nature; it alone did not disappoint (1834:113).

A favourite walk for visitors and natives was several kilometres in length, northward from the square, and up to the rim of Cerro Macuiltepec, a cinder cone. Three hundred and sixty degrees of view: the sea in the distance to the east, Mount Orizaba to the south, the Cofre de Perote to the west, a rugged chain of mountains to the north, and the town itself just below. Vistas on a more intimate scale could be found near a lake in a valley just south of the town. Better still, Becher noted, was the new road leading southward toward Córdoba (1834:111). This was the favourite promenade of Jalapa's beau monde. It led over a pretty bridge that spanned a waterfall, through forests and some cultivated lands to the small town of Coatepec. All this was an idyll, complete with sombre dells, gurgling rivulets, white-washed cottages, and sugar plantations (Robertson 1853:282–3). Jilotepec, a village to the north of Jalapa, was a similar destination. One descended to it dramatically; there it was, a gem in its setting. The view had to be "a noble blending of the sublime and beautiful" (Robertson 1853:280).

Bullock recounts a picnic excursion in the company of some twenty people to a sugar hacienda, probably La Orduña, just east of Coatepec (1971:58-9). The group was first taken through a brandy distillery, then they were invited to a rustic pavilion where a plentiful repast was served, including a whole roasted three-month old pig, stuffed with walnuts. During the meal a strange game began. Someone rolled a piece of bread together into a ball the size of a pea and flung it surreptitiously into the face of another guest. The favor was returned until the battle became general. Bullock seems to have been taken aback: both ladies and gentlemen were contending, whole cakes were being expended. Then the battle moved from the pavilion out into an open space. One is reminded of filmed summer decadence, all bizarre angles and slow motion. Anything that could be thrown was thrown: "and when every thing else had been exhausted, what had been left by a number of mules that had been feeding by, was sent about in all directions by the combatants." There was leapfrog then, until the sun was low, when everyone returned to the house. The final scene, Bullock was given to understand, was the typical conclusion of all Spanish parties:

Cards were produced, the table was immediately spread with doubloons and dollars, and considerable sums were won and lost in a few minutes. I was shocked to observe the change which took place, and in so short a time, from boisterous but innocent mirth, to a display of passions of the worst kind, and in which the ladies acted a still more unpleasant part than in the former sports. Those beautiful beaming black eyes which, but a few minutes before, had sparkled with life and joy, were now overcast and lowering with expressions of avarice and discord: not one smile nor jest occurred during the whole of this short scene. . . . It was some time before hilarity resumed its sway, when some curvetting and racing took place among the sprightly little chargers. We entered the city of Xalapa after sun-set. . . . (1971:62-3).

CHAPTER TEN

Leaving Enchanted Ground

Curious Mexicans had lined the wharf at Veracruz to watch the foreign visitors land; they had eyed them at many points along the road, and they might well be there again in Jalapa, on both sides of the coach, watching the preparations and then the departure for the pass.

The travellers were treated to a few kilometres more of gratifying landscape: trees from their temperate homelands, dairies, and cultivated fields. Even today, after the last suburb and the cement plant, one can still sense a bit of that pastoral. A sign says "Rancho Liquidambares"; another, "Rancho Paraíso." Urbanites with means have long built country residences here.

Then everything changed. Ward, for one, was amazed at the sharp upper boundary of this mid-slope paradise: "Its beauty vanishes the instant you wander beyond the limits of the enchanted ground. . . . [We were] prepared to expect a great change in temperature, although we were far from calculating upon a transition so sudden and so complete, as that which we experienced" (1828,II:194).

With the colder temperatures many travellers also encountered fog. It could set in quite soon after Jalapa or later, nearer to the pass. It darkened and lightened, wrapping veils around the travellers, passing among the trees like phantasmas (de Fossey in Glantz 1964:256). This is common during the *nortes* of the dry season: it also occurs during the wet season (Figure 3), when moist tropical air is driven westward up the slopes.

The travellers were moving out of the tropical world; the indications around them were unmistakable. They were now to add several facets more to the landscape they had fashioned together in their accounts, to elaborate on its resources and its hazards and indeed the mood of it up here at its upper margins; they were also to give their assessment of its people a final twist. Before they went over the summit they would be given a visual stimulus for a summary of what they had seen so far. Several of them attempted an integration by means of what is still quite a useful construct.

The great variety of the mesophytic natural vegetation in the environs of Jalapa diminishes upslope. The oaks that signal amelioration east of the town and continue as one of the many genera in the vegetation of paradise are also one of the two dominant elements in the forest above it. Conifers are the other. Both are adapted to soils developed on volcanic parent materials, such as those that cover the skirts of El Cofre de Perote (Figure 1). They are so intricately related in their distribution and succession that they are often used jointly to designate one forest type (Rzedowski 1978:284). As one nears the summit of this particular pass the oaks diminish and the forest becomes a solid stand of conifers.

Sartorius described the slopes above Jalapa well. He differentiated among the conifers, noted the grasses, clovers, and wild flowers. He found slopes were cut at intervals by ravines, with foaming torrents. Among the pines and firs there was stillness, punctuated only by the screams of jays and the occasional howl of a wolf. This was a landscape lonelier than the open grasslands. Here and there he noted meadows too, which stimulated the expression of a comparison already implied. Here were "all the charms of Alpine vegetation. All is familiar to us" (1858/1961:16). This was a Germanic forest; the melancholy it evoked was different from the amazement and unease one felt at the edge of the tropical forest down in the humid lowlands.

The ascent through the upper slopes of the escarpment could be a miserable experience, as it was for the men of various American army units who marched upslope after the battle of Cerro Gordo (Smith and Judah 1968:224; Kenly 1873:364; Brackett 1854:72-3). During the day it could be quite warm, but the nights were cold and the tents dripped with humidity. When a *norte* struck, there was streaming rain, or worse, hail and snow, and limited visibility, which was disconcerting since the men were still subject to guerrilla harassment.

Melancholy will have settled into many a coach, as reported when Ward made the ascent (1828:194-5), and again when Mayer was coming up (1844:18-19). The civilian travellers seem turned in on themselves and, with some exceptions, as unresponsive to the pine and oak forest as they had been to the chaparral. There is nothing like the enthusiastic characterization of the tropical forest in the humid lowlands or the mesophytic forest around Jalapa. The commander of the French Expedition of 1838, Admiral Baudin, when on his way to negotiations in Mexico City, was impressed by the economic potential in the conifers. He was mentally stripping them of their greenery and putting sails on them (Blanchard 1839:119). This resource had been one of the arguments during the colonial period for a shipyard somewhere in the lowlands, but nothing ever came of it, and the ships were built in Havana instead (Siemens and Brinckmann 1976).

One feature of the physical landscape did usually arrest the travellers' attention: the lava flow that stretches across the road between La Joya and Las Vigas (Figure 20). It was and still is called *malpaís*, literally *badland*, which is generally applied to such terrain in Mexico. Contorted dark shapes lay all around; the travellers swept their gaze back and forth across them, some checked their arms. Here and there one saw the remains of fortresses used in various campaigns—intended for defence against the Americans, too, but never used. Jamieson, marching upslope with his men, was moved: "Pine, cypress and hemlock trees of gigantic size have sprung up in the midst of its desolation, and cast a melancholy shade over the ruins around. Their evergreen branches wave and moan over the black and ashy surface beneath, in silent and solemn contract—one an emblem of life immortal and the other of death everlasting" (1849:41).

Tudor believed this landscape surpassed in savage wildness the defiles of the Italian Alps and the Apennines. It was an ideal haunt of *banditti*, which must have been as much a reference to the art of Sal-

20 Malpaís

vator Rosa as to the more prosaic likelihood of an actual robbery (1834:188–9). He saw crosses erected along the road and interpreted them to commemorate murders committed during robbery. It seems this was largely so, but they may have indicated other sorts of fatalities too, such as those occurring in fights between *arrieros* (Trens 1955:44). Many crosses may be seen today along the highway between the summit and Jalapa; they represent the toll of accidents.

As to human habitation between Jalapa and the pass, the travellers noted several changes in the construction of the houses. The white walls of the town and of the settlements in its vicinity gave way at about the altitude of San Miguel to dark walls of volcanic rock. Above Las Vigas most of the houses were walled with planks and roofed with shingles. These reminded Ward of houses he had seen in Sweden (1828, II:195). Bullock remembered mountain villages of Norway and the Alps (1825:67).

These variations in the construction of the common Mexican dwelling represented the local availability of certain building materials and perhaps also post-contact stylistic introductions. They are an excerpt out of Mexico's rich settlement geography. The wooden house at the altitude of the conifers has always been one of its most striking features.

And, as to the inhabitants themselves, there were curious changes with altitude as well. Humboldt pointed out that not far above Jalapa the banana no longer ripened. The plant that more than any other symbolized the scandal of easy tropical life was not available. A more vigorous environment awakened industry and constrained the mountain dwellers to work (1810, Band II:173). Mayer found them in fact to be hardier and more robust than the people downslope (1844:19). De Fossey was convinced that the songs and dances of the *jarochos* ceased above Jalapa. People lost their joyfulness and became introverts (in Glantz 1964:255).

Those travellers fortunate enough to come along this mountain road above San Miguel when most of it was clear of fog were treated to a fine last view across the lowlands. One traveller found it repaid him for all his fatigue (Tudor 1834:177–8).

To the northeast Bayard Taylor, the journalist returning from California, saw that "the beds of the valleys, wild, broken and buried in a wilderness but little visited, were lost in the dense air, which filled them like a vapour" (Taylor 1850:324). These valleys were indeed well off to one side of a main route, seen by passing travellers, but little visited, tantalizing but remote, as they remain for the passing traveller to this day. There are many details of Central Veracruzan history yet to be sought out within them and recorded. The whole region is sown with the remains of Prehispanic occupance; its valleys were historic refuges and offered quite attractive bottom lands for agriculture too.

Turning somewhat toward the east the same observer reached for a metaphor: "Fancy yourself riding along the ramparts of a fortress 10,000 feet in height, with all the climates of the earth spread out below you, zone lying beyond zone, and the whole bounded at the furthest horizon to which vision can reach, by the illimitable sea!" (Taylor 1850:324).

Here at the ascent out of the lowlands, as earlier at the approach, and at many points in between, there were echos of Humboldt. The observers were seeing what they had been led by him to expect they would see. He had outlined the altitudinal zonation he had found in the Andes and now followed through on these tropical slopes as well. The snow fields at the top of Mount Orizaba were at the one extreme of an interdigitated sequence of life zones and the forest around the wetlands just behind the dunes was at the other. It seemed to Humboldt that: "The admirable order with which different tribes of vegetables rise above one another by strata, as it were, is nowhere more perceptible than in ascending from the port of Vera Cruz to the table land of Perote. We see there the physiognomy of the country, the aspect of the sky, the form of plants, the figures of animals, the manners of the inhabitants, and the kind of cultivation followed by them, assume a different appearance at every step of our progress" (Humboldt 1811, III:251).

Robert Wilson, the travelling lawyer with a strong turn of phrase, gives the altitudinal scheme a flourish. He was riding on the top of the coach and the view, he maintained, was such as the devil had used to tempt Christ, taking him up into a high mountain and showing him all the kingdoms of the world in a moment of time:

This changing vegetation is a barometer which, in Mexico marks the ascent and descent as regularly as the most nicely-adjusted artificial instrument. So accurately are the strata of vegetation adjusted to the stratas of the atmosphere which they inhabit, as to lead the traveller to imagine that a gardener's hand had laid out the different fields which here rise one above another upon the side of the mountain that constitutes the eastern enclosure of the tableland. . . . This specimen chart, where all the climates and productions of the world are embraced within the scope of a single glance. (1855:80–1)

Eventually fog swirled around Wilson too, as it comes to most travellers going over the pass, whatever the month may be. And with the fog came a chill and piercing air; the pleasure of their mountain ride was over.

One hears a diminishing clatter of hooves and wheels on the paving stones, and then there is this note from Ward. When his party was over the pass and out on the tableland they found "little to attract the attention or gratify the sight" (1828:199).

CHAPTER ELEVEN

Round Unvarnished Tales?

Left standing on the ramparts, the reader may well want to turn back to the view, which on a good day is clear all the way to the sea, and reconsider what was outlined and shaded by the observers' accounts— in the light of their own intentions, expressed or implied, as well as from the perspective of our time. This can easily become a reflection on facts and fiction, on reasonably firm data and conjury.

SEEING THE LAND, AND ITS PEOPLE

Poinsett's account, the earliest on our list, freshens the exhilaration many felt on their approach to Veracruz from the sea:

> As the sun rose, we saw the land, and a magnificent sight it was. The outline of the rugged mountains that skirt the Orizaba was clear and distinctly defined; and the clouds that hovered round their base and obscured the low lands near the coast, were in motion and rising gradually. The summit of Orizaba, a regular line covered with snow, towered high above the loftiest mountains. We continued to gaze with delight on this view until ten o'clock, when the whole was enveloped by the clouds, which rose and uncovered to our view the low lands. (Poinsett 1824:11–12)

Landscape descriptions tend to foreshorten, of course, as do paintings and drawings; this tendency is often pointed out and easily verifiable. The travellers approaching Veracruz distended a landscape that will have been fairly low on the horizon. After the long voyage they were more than ready for land. And many noted that it was banded, like the even grander vision that would confront them just before the summit on departure. A line of dunes made up the base, with the city embedded in them, a cutout of domes, towers, and low rectangular shapes that grew as one approached. Behind this was the promising ribbon of green, within which would be the tropical luxuriance of which one had read, and which was now of special fascination: "Undulating, finely diversified with hill and dale, and rivers and forests, and rich in the deep verdure and luxuriant vegetation of the tropics" (Poinsett 1824:12). It faded on the higher slopes of the mountain range, into the snow fields on the dormant volcano.

Various fears edged the descriptions. There had already been the danger of shipwreck in the Gulf and there would be bandits along the road into the interior, but the worst threat was death by yellow fever; it would hang over the travellers until they reached the oaks below Jalapa, but would become especially strong while they were in the port.

Before the travellers could come ashore there were often delays, during which there was little else to do but scan again and again the fort of San Juan de Ulúa, the Isla Sacrificios, and the waterfront of Veracruz. They appear to us in disproportionate detail. On landing, the scale of the observations increased, everything was now close up, and the pace of the narratives speed up as well. The reader feels jostled on the wharf, with just a moment to appreciate the tableau of Veracruzan spectators sketched by Madame Calderón de la Barca or Sartorius, and then is surrounded by movement and noise in the streets. What was seen and heard is vivid, but the sounds seem inarticulate: few of the visitors knew any Spanish.

Visitors from North America and northern Europe found the structure of the town, both its layout and what they saw of the interiors of the better houses, uncongenial—a clear example of a dissonance in urban traditions. Groups of foreigners engaged in trade were noticed with disapproval at tables in questionable places, sallow-faced, drinking too much, and making too much noise. That was what happened to people interested enough in money to stay in a place like this. Most visitors were impatient to leave, but a few did have the peace of mind to appreciate the entertainments of Veracruz and have left us some fine impressions, such as the ladies on the plaza of an evening, wearing fireflies, or travellers, safe on land, watching the clouds and waves of a *norte* through the watergate.

While detained in the town the travellers struggled with thoughts about yellow fever, its cause, its trajectory, and how it might be ameliorated. In the light of what has been learned of it since, reading of that horror now is like the sighted watching the blind. Most observers were convinced, of course, that the disease emanated from just beyond the walls, in the stagnant water impounded within the dunes, and in the wetlands further inland. It induced a feverish view of a town that was seen to be lively, clean, or at least being cleaned, that was orderly and indeed painted in attractive hues, but known to be deadly. Such a town epitomized the riskiness, and the attraction, of the tropical lowlands. One expected excitement but worried too that one might be punished for having ventured into the lowlands. It was important to maintain one's equanimity, one's good habits, and to get out of town as quickly as possible.

In their departure, many travellers betrayed a certain truculence toward what they might encounter along the road and indeed toward the country as a whole. Let any inhabitant attempt any sort of nonsense and they would be answered with gunfire. One didn't need to worry too much about return fire, either, because Mexicans were cowards and couldn't shoot straight.

Just east of the dunes the travellers finally came on that green band they had seen from the decks of their ships. They expected the full luxuriance of tropical nature, and it was there, a richly varied greenery, not in vast expanses as other travellers would find it in South America, but as a sample one could pass through in a few hours. The naturalists in our company were enthralled; others appreciated the richness, too, but were not comfortable with it. Henri Rousseau grasped this well in his paintings, which have come to exemplify "jungle" to this day. There was menace in such a forest; a visitor from temperate lands was not much inclined to walk in these woods. The very exuberance became a part of the indictment of the lowlands; the luxuriance was unbridled and it gave cover to malicious plants and vipers. Here and there within it lay the swamps, the sinks of putrefaction.

The United States army that landed south of Veracruz in 1847 had to fight, as it happened, in the dunes just behind the beaches and in the dryer hill lands to the west of the low-lying terrain with its forest and swamps. Their accounts are mostly disappointing in their landscape descriptions, but they do approach eloquence in what is said of the heat, the sand, the insects, and especially the low thorny vegetation covering the fixed dunes as well as the hill land, vegetation which they called chaparral. It was luxuriant in its own hideous way and it hid the guerrilla, who was really just a bandit and who fought in a manner that was uncongenial, unmanly, unfair, and very Mexican. As the soldiers wrote home and then personally told their war stories, the guerrilla, and indirectly also the chaparral, were made a further part of the nineteenth-century North American indictment of the country.

Thus, before the observers had sensed the amelioration that would come at about 1,000 metres they had already rung several changes on the luxuriance of tropical nature. By this time too they had scattered many comments on the people of the lowlands through their accounts.

These were not so carefully or enthusiastically observed as the physical environment, to be sure, and the descriptions were certainly not very impressive ethnography. In fact what is said now strikes us as anything else than realistic. It seems facile and prejudicial. The Indians lead an "easy lazy life" (Tylor 1861:315), the *jarocho* was a villain and the low-lander in general was a "cumberer of the ground" (Tudor 1834:178).

The subculture of the *jarocho* is outlined in some detail. It was a term used for the common lowlanders of various occupations: herders, peasants, fisherfolk, even the poorer townspeople. It did not include the creoles nor the Indians and certainly not the foreign born; it was a term for people of mixed racial background. It had opprobrium embedded in it and was often used by uplanders in deprecation of lowlanders, who might easily turn it around and make it a matter of pride, as they do to this day.

The opprobrium was applied much more to the *jarocho* than the *jarocha*, in fact she was often sketched in quite seductive lines, with special appreciation for her loose clothing. Most of the authors were men and some of them could hardly hide their lechery. Fanny Calderón de la Barca was scathing in her remarks on the *jarocha* and other Mexican women too, and most especially the women of families with means.

The *jarocho* was nonchalant, he pushed his log whole through the door and fed it gradually into the fire rather than cutting it up beforehand. He was clearly improvident, and temperamental, also untrustworthy, cowardly, and lazy. His material goods were laughable, the typical house was little more than a slatted chicken coop. Ranching was suitable to his lack of industry; there was little evidence anywhere of the plow, and this was basic.

Mount Orizaba, whether veiled or resplendent, had been a strong signal that focused the attention of incoming travellers and engages the reader of their accounts in an appreciation of place. The first oaks below Jalapa were signals too—of a more immediate and pragmatic significance. Here the temperature dropped a little and the danger of infection with yellow fever faded. One could breathe easier—going up, or going down. The chilly warm air warmed for the traveller descending the slope, an amelioration not quite so vital, perhaps, but one that lifted the spirit and loosened the tongue.

Above the oaks the travellers could observe in a more leisurely way; the suspicion that there was danger in the greenery fades from the accounts. Here was luxuriance indeed, exotica that could be enjoyed in comfort. Here the rendition of the landscape climaxed. All this taxed the visitors' prose and inspired those who could draw or paint. The rhapsodies now seem excessive. One is amused, rather than surprised, to read that the inhabitants of Central Veracruz were seen to improve in appearance as the travellers entered the favoured stratum: these were unguarded observations indeed! However, these people were not all that impressive either. The challenge of this particularly promising terrain had not been met any more successfully than down below. The travellers still saw very little agriculture.

For those departing upslope, paradise ended not far out of Jalapa. Temperatures came down and fog or rain set in. The great variety of vegetation diminished until the pines predominated. The trees, the fog, and the volcanic debris of the *malpaís* imparted a few last romantic shivers. Above it the forest looked familiar. The house in the villages were more and more of wood, the people heartier still and more robust, but less joyful too. Then came the view.

YEARNING FOR FACTS

Various of the travellers in Ortega y Medina's theatre troupe affirmed that they intended to write of what they had seen or personally knew to be true. This aspiration was particularly strong among the naturalists of the late eighteenth and early nineteenth centuries, as Barbara Stafford has developed in her *Voyage into Substance* (1984), but it can be detected more broadly within our troupe. Their readers, they maintained again and again, were to expect good factual information. Their observations were often published as originally entered into their journals or were heavily reliant on such material from other sources. Moreover, there were few reasons for most of the authors to dissemble regarding what they saw in the tropical lowlands; this region was seen en route to places of main concern: the capital, the mines, California, Acapulco, or if they were descending to the coast, the port of Veracruz itself. One could be frank and unselfconscious. It could equally well be expected, therefore, that predispositions would show through particularly well. It is useful, however, to weigh up the factual yield first.

One can abstract from the foreign observers' accounts a good deal of the physical form of Veracruz and Jalapa, as shown in Arreola's essay on the nineteenth-century townscapes of eastern Mexico (1982). The structure doesn't change much during the half century under consideration, except that warfare leaves its destruction. The observers were not very helpful with their population figures; these rise and fall with little plausible explanation, but this is not surprising, given the conditions of the time. Koppe's careful citation of official materials made available in the early 1830s is the exception. From time to time many of its inhabitants had to flee and visitors arrived on what has the feel of an empty film set, but eventually the people would come back again. Total population figures dipped generally during the half century, but then came up again to pre-Independence levels.

There are only intimations of the desperate living conditions that obtained for most of the townspeople of both Veracruz and Jalapa, and the other smaller settlements too, for that matter, during much of the first half of the nineteenth century. Had these been more strongly sensed the rendition of this landscape and its inhabitants would surely have been less censorious. There is some useful detail on abandoned housing in Veracruz, as well as the terribly deteriorated road. The impact of siege and bombardments on Veracruz is graphic, as is the blight left on the villages by the passage of contending insurgents and an invading army. The most telling, and backhanded, indications of local desperation arise from the discussion of highway robbery. Various Mexican authors of the time made the difficult conditions a prominent theme in their consideration of the region (Rivera Cambas 1959–60; Trens 1950).

Imagined towns obtrude in the travellers' accounts. Veracruz was first and last an ominous place of death, which no amount of evidence about clean streets and bright paint could ellipse. Jalapa was the languorous maiden spread out at the approach to the escarpment, praised without inhibition, as she is to this day. The wet and miserable foggy days of her winter could not diminish her.

Transport and trade through these towns and across this lowland landscape could be expected to absorb a good deal of attention; many observers were looking for investment opportunities. The movement of goods from Veracruz into the interior was particularly fascinating: what might be expected to happen to machines, for example, if they were shipped in by muletrain and it rained en route. The movement in the other direction was found surprisingly limited, and from that, of course, one could deduce much of the country's economic predicament. There is in fact substantial information about the organization, function, fluctuation, and difficulties of the commerce in imports and exports. This has been drawn on, as in Pferdekamp's analysis of the development and decline of the *almacén* (1958:55–66), or Herrera Canales' *El Comercio Exterior de México: 1821–1875* (1977). Peter Rees has shown the general state of inertia in the development of transport and commerce between Mexico and Veracruz from 1519 and 1910, as well as the competition between routes (1976). It is apparent from a number of the accounts, moreover, that local commerce in daily essentials persisted remarkably in spite of unrest, although there are indications that the travellers were responding more to exotica than abundance.

Some affirmations on the phantasmagoria of yellow fever turned out to be well-based. The disease is altitudinally limited, but those aspects of the vector's physiology that are responsible for this are evidently not yet well understood. Lowlanders did achieve immunity in slight attacks during childhood, as is hinted. All of what is said about this disease is set off, as it happens, by some good information on illness and morbidity among the population at large, rather than just among the unacclimated foreigners. The worst killers in the lowlands were various pulmonary and gastroenteritic diseases.

The actual circumstances of travel and the conditions of the road are recounted in some detail, but much of this is probably not unvarnished either. After a certain number of descriptions of inns, restaurant fare, coach travel, and suspicious-looking escorts, one comes to appreciate the witty variations: the poor man who had to ride the stagecoach seated on his pistols because there was nowhere else to put them, the guest who found to his considerable amazement good food and a quite respectable wine at an establishment in Plan del Río, the young men of the first English diplomatic mission who surveyed their party disposed for the night in and around a shed in Santa Fé and chuckled over the hesitation older members of His Majesty's foreign service would show when asked to take up posts in one of the new Spanish-American states.

The road was undoubtedly insecure; to travel it meant risking robbery. But to arm oneself might be most unwise, as the guns of a foreigner could easily invite violence and endanger everyone in the coach. It was more adaptive to think of banditry as a toll. There were subtleties to be learned here, but most travellers never noticed. Instead, the insecurity, the condition of the roadbed, and many other aspects of travel in this country at this time were made elements of the indictment.

A particularly frustrating aspect of this literature is the thinness of the material regarding the use of the land. What there is, is given incidentally, and amounts to little more than a few observations on skimpy fields and unkempt pastures, except in the case of Sartorius' discussion of ranching practices in the lowlands and the engravings by Rugendas that enhance his text.

The modern reader can hardly avoid reflecting on what was missed or misread, or perhaps one should say, what was not yet apparent to the outside observer. The luxuriance of the tropical vegetation and the potential of these lowlands for agriculture was overestimated, but that has been a problem in lowland and development planning until very recent times. The actual fragility of this environment is still inconceivable. The difficulties of making a living, of winning from the tropical environment even the modest amounts of agricultural produce noted by our travellers, were underestimated.

There is careful commentary in Mexican sources, and in several of the more systematic and incisive foreign sources too, on the unfortunate

tenure patterns of the day, the deficient infrastructure, and the unintegrated economy of the lowlands and indeed the country as a whole.
There were reasons why it made little sense to work but these were not
recognized by most of the foreign observers. They were in fact seeing
sensible subsistence strategies, given the political and economic circumstances of the time. Sartorius went so far as to acknowledge a certain
insouciance among rural Veracruzanos, but that was little more than an
isolated ambiguity in the general verdict of improvidence. The landholders of substance were indicted too—for their lack of entrepreneur-
ship. They were in fact pursuing gain by the few means open to them.

The judgment of the inadequacy of the people of the tropical lowlands, diffused and almost standardized in North America and in
Europe, is only gradually being expunged in our time. To an articulate
jarocho like the Veracruzan author Melgarejo, such a view of his countrymen has been infuriating, as we have seen.

The vegetation was the most immediately apparent physical aspect of
this landscape. Except for Sartorius' altitudinally conceptualized and
richly suggestive descriptions, we are presented mainly with more or
less elaborate lists of species. Schiede's extended observations are really
only an interestingly annotated enumeration. Advances in taxonomy
and much new data have rendered many of these notations curiosities.
What the various communities of plants suggested to the viewers was
another matter: the luxuriant, if unsettling, tropical lowland forest
proper, the frightening swamps, the dismal dryer forest on the hills a
little farther inland, the dense and thorny chaparral through which the
soldiers had to make their way, the sentinel oaks below Jalapa, the
hardly describable vegetation around the town itself and then the dark-
ening oaks and pines below the pass.

The descriptions give us a good deal of the rich variety and exuberance that has become a commonplace knowledge about tropical lowland rainforest. We are fortunate to find several detailed considerations:
a riverbank, a transect through a part of the San Juan Basin, and a fine
menacing swamp. Considerable stands of palms are described which
indicate repeated applications of fire over long periods of time in aid of
cultivation and ranching. Cattle are noted leaving the wetlands proper
with the onset of the rains, moving into the newly green neighbouring
hill land, and coming back down again into the wetlands during the dry
season—clear evidence of the persistence of tropical transhumance.

As might perhaps be expected, there was little information on landforms, geological features—except in the *malpaís*—or soils. Weather was
dealt with at some length in many accounts; it is described almost as
specifically as the vegetation. With respect to the *norte* and weather

phenomena in the favoured zone, some authors, especially those of the systematic sources, synthesize sufficiently to give us creditable and still quite useful outlines of climatic regimes.

The most interesting aspect of the environmental observations is the altitudinal zonation that is inferred. Subsequent observations along other tropical slopes in the Americas have elaborated on the integrative potential of the construct. The investigations undertaken in the Andes early in the twentieth century by the German geographer Carl Troll stand out (1959). They were conceived as a fleshing out of the Humboldtian scheme. Troll's profiles, maps, and comparative tables are some of the most useful visual materials on the physical environment of the region. The anthropologist John Murra has outlined a cultural-historical articulation with altitudinal zonation in the same region (1956/1978). His concept of the role of verticality in Andean economic and political life has been found highly plausible (e.g., Forman 1976). This paradigm has not yet had as wide a currency in Mesoamerica but here and there it has been found of use. Sanders and Price, for example, allude to altitudinal zonation in their ecological regionalization of Mesoamerica and emphasize the importance of interregional symbiosis in the evolution of its cultures (1968).

Various specific environmental studies have addressed altitudinal variation: the study of temperate zones on the slopes of the Sierra Madre Oriental, already mentioned as an approximation of a geography of euphoria, is an excellent example (Miranda and Sharp 1950). Many investigations of the Instituto de Investigaciones sobre Recursos Bióticos in Jalapa might be cited as well, such as the description of the vegetation of the environs of Jalapa by Zolá Báez (1980).

Recent work on the Prehispanic agriculture of wetlands in Central Veracruz has clarified and assessed the cultural-historical significance of micro-altitudinal complexities at the base of the region's altitudinal profile. The patterning in the wetlands and the newly plowed fields on the sloping tablelands just west of the band of wetlands suggest an elaborate altitudinally articulated subsistence system (Siemens 1983). And any of the many barrancas that slice this sloping terrain indicate a transverse natural and often agricultural zonation along its sides that further complicates the present picture, as well as what may be envisaged for the distant past. It would seem, therefore, that it is in the productive potential of linked micro-environments that the real luxuriance of the tropical lowland lies.

In their bold restatements of the Humboldtian scheme the nineteenth-century observers were verbalizing something like a research objective that would eventually be followed up by serious scholarship, but they were also oversimplifying the natural order. It is difficult to accept the sweeping "specimen chart" seen from the ramparts; it was more imagination than observation.

SEEKERS AFTER THE PICTURESQUE

Nineteenth-century observers have been strongly dichotomized: "I am at pains to distinguish and dissect the scientific aesthetic of discovery from that concurrently espoused by the followers of the Grand Tour or the mere seekers after Picturesque scenery . . . a genre whose very popularity is based on ardent yearning for facts rather than fictions" (Stafford 1984:xix, xx).

Stafford's exposition is quite absorbing, and to a considerable extent its main findings regarding the "scientific aesthetic" are recognizable in the accounts that we have analysed, particularly but not exclusively in those of the naturalists. Facts and the pleasing picture (a fiction?) were often intertwined aspirations, the first likely to be explicitly affirmed, the second implied by the actual content of the accounts. It is the imaginative colouring that often provides the more interesting, diagnostic reading, and much of the best of that is provided by the accounts of the naturalists. This is the "data" in the accounts that has been least made use of. And then there is a further sense in which fiction seems to become fact. The way in which tropical nature and tropical people were shown by nineteenth-century observers surfaced eventually as accepted wisdom, overshadowing perception and quite apparently decision-making too, well into the second half of our century.

The imagery with which the travellers coloured the setting through which they made their procession came from all the expectable sources. Classical allusions were frequent: the *norte* became the dragon guarding the golden apples of the Hesperides; Puente Nacional became Thermopylae. Paradise was given various classical designations, as well as those taken from the Bible. Hell was too strong a term for the *chaparral*, it seems, but purgatory was drawn from Roman Catholic doctrine and pressed into use. Near Eastern comparisons came easily, not only because that region was the setting of biblical stories, but also because it had long been a principal source of exotica for Europeans. Some travellers had been there before coming to the New World. Desert images were used for the environs of Veracruz. The Moorish element was evoked. Andalucia was referred to in descriptions of Jalapa. The best way to make Mexico understandable, it seems, was to make circum-Mediterranean comparisons. Several authors called Mexico the Italy of the New World.

A shadow was drawn over the lowland landscape and indeed all of

Mexico, as is well known, by what is commonly termed the *Black Legend*. This was a profound and widespread antipathy, directed first from North Europe to Latin peoples and especially Spaniards, then by extension from Europe and North America to Latin America. It was especially strongly held by those travellers who were of Protestant background. The main particulars of the indictment were cruelty in the treatment of indigenous peoples, laziness, impracticality, bad faith, political corruption, and even uncleanliness (Gibson 1971:179). Most travellers, before they had spent many days in Mexico, recognized most of these characteristics and readily traced their effects.

Their accounts do allow us, however, to appreciate a series of distinctions which are not so common in the literature on the animadversion. They often sketched out a kind of human altitudinal zonation: the people of the tropics were most villainous in the lowlands proper and improved with altitude. But this hardly applied to Indians, who were regarded as relatively harmless and not culpable in the main for the unfortunate conditions that prevailed. However, little could be expected of them, either, toward future economic and cultural development. The lowlander of mixed racial background, the *jarocho*, he was the main culprit. And the gender-specific pronoun is of key importance. In the Mexican lowlands, as elsewhere, the woman was admirable; all things considered, it was the man who was truly reprehensible.

In their descriptions, the travellers often seem drawn back and forth between fears and attractions. Shipwreck, yellow fever, and robbery threatened them in succession. The first two were awesome and humbling, the third invited resistance. One needed always to remain on guard, not only against these specific threats but also against the insidious possibility of relaxation, the slackening of one's system and one's principles that could occur in the tropics. Added to this was the daily need to deal with countless bothersome practicalities. All this tended to make of the crossing of the tropical lowlands the first leg of a Mexican obstacle course, an ordeal, a kind of secularized Pilgrim's Progress.

At the same time this New World was intensely attractive. It promised to satisfy old longings for the warm breezes and the oranges of the south. Most visitors expected some adventure and entertainment. Those with a naturalist's bent were determined to gather a rich harvest of strange new information. What might therefore, on the one hand, be hateful, bothersome, and dangerous, could be, on the other hand, quite fascinating, all of which led to some intense prose.

Eighteenth-century travellers have been described as in search of the picturesque landscape: wild, primeval, varied, without order, and full of contrasts (Immerwahr 1972:28). Salvador Rosa's paintings gave them their standards. Our travellers of the nineteenth century were certainly still sensitive to such stimuli. They appreciated seascapes roiled by *nortes*. Chasms were a thrill, preferably stocked with bandits, as was tropical vegetation, which some construed as an English garden. They noticed ruined houses in Veracruz, overgrown with greenery, although they didn't fully credit the desperation that these represented. They were mesmerized by Mount Orizaba and the layering of environments visible on its slopes.

The picturesque was found on a larger scale as well, in the details of this tropical place. Madame Calderón de la Barca was convinced that in Mexico, "there is not one human being or passing object to be seen that is not in itself a picture . . ." (Fisher ed. 1966:69). Many of the rapid verbal sketches in the literature reviewed were skilfully done and remain arresting: a topless *mestizo* girl, glimpsed in a doorway, taken to symbolize the seductive beauty of the entire New World (de Fossey in Glantz 1964:254), families grouped in front of their houses for the few moments it took a conveyance to pass by—a touching picture of patience and poverty (Kollonitz 1867:76). Several authors give us especially striking glimpses of nocturnal fiestas at coach stops (e.g., Vigneaux in Pasquel 1979:249–50).

In a great deal of what we are shown from horseback, out of the coach window, or, with luck, from the seats up on top, one recognizes aspects of Romanticism: a "maddening subject" as one of its chief exponents has declared (Peckham 1970:28). Indeed, some of the most influential residue of nineteenth-century observations on tropical Mexico derives at least partly from this sensibility. Romanticism is a great many different things. It cannot easily be reduced to a set of key tenets, against which one might test a given piece of literature, although that has been tried (Jones 1961; Praz 1962; Meir 1963; Peckham 1970:66–70). We can make some sense of it from our company of authors and the landscape they left us if we narrow our attention to what might be called an early Romanticism. This is a congeries of attitudes that emerged from the Enlightenment, and found strong expression in the literature of the eighteenth century. Most of our travellers were neither poets nor artists in prose; they had themselves disclaimed any such aspirations in many a preface. The literary influence, however, can be detected and the allusions are there. Romanticism has been called a literary emotion; "it is Nature seen through the medium of literature, through a mist of associations and sentiments derived from poetry and fiction" (Logan P. Smith, in Immerwahr 1972:12). The actual attitudes of which the Romantic tendency is composed may be designated in various ways: love of nature, melancholy, enthusiasm, or sentiment. Morse Peckham has maintained that these were means "to discharge the tension consequent upon the affirmation that the world is radiant with order and value . . . and the inconsistent perceptions that it is not" (1970:30–1). Regarding the nature of that "order," we may perhaps resort to Loren Eisely's formulation: "In the pre-Darwinian portion of the nineteenth century, we encounter what is really a combination of traditional Christianity overlaid by a wash of German romantic philosophy. Elements of the new science and the new discoveries are being fitted into what is regarded as the 'foreordained design of the Creator' " (1961:95).

Contradictions were suspected in tropical nature itself; the ceaseless battle of life, apparent especially in the forests and wetlands of the humid lowlands, the dangers that lay in wait for the unwary, filled many a traveller with unease. However, the dissonance between relatively "orderly" nature and quite disorderly human activity was clear.

INDICTMENT

Humboldt, who has been called both an Enlightenment figure and a Romantic (Miranda 1962:138), rendered the key dictum. In the altitudinal zonation between Perote and the sea he saw an "admirable order" (Humboldt, 1811, III:251). This was no isolated formulation. He had written of this earlier while journeying in South America and was eventually to find in it much of the basis for a new discipline: plant geography (Troll 1959). In the works of man along these same slopes, on the other hand, he saw inadequacy and disarray; most of the travellers that followed took his direction. In their various ways, with greater or lesser degrees of detachment and delicacy, they expressed their amazement over the luxuriance of tropical nature and the failure of its inhabitants to resist its blandishments and make proper use of it. This is the bitter aftertaste of the nineteenth-century observations.

The indictment was often broadened to include the whole country; it became part of the long bill of particulars held against Mexico in North America and in Europe, on which a good deal has been written (e.g., von Mentz de Boege 1980:146; Robinson 1977). One can see the critique being elaborated and given facets not often noted as our observers journey into the interior along the Veracruz-Jalapa road: a fine volcanic peak, rich greenery in the lowlands, dangerous for its thorns, the biting ants it hid, and the swamps with which it was interspersed, but luxuriant nevertheless, the greenery at mid-slope unreservedly luxuriant, in which the Creator had outdone Himself—all this outweighed by something one might see already in the faces of the pilots, a certain shiftlessness, fully apparent among the half-naked *canaille* of Veracruz itself, as well as vacuously docile Indians in their tableaux, insouciant *jarochos*, who could hardly be taken seriously, and the despicable guerrilla, who was really a bandit: piece after piece added to the mosaic of disrepute.

The indictment had an authoritative base and was widely diffused. Humboldt was adulated in Mexico and abroad. His *Political Essay on the Kingdom of New Spain*, and his various other works as well, were seen as models of disinterested analysis. Information on Mexico was avidly sought in North America and Europe; the travellers were often under pressure to publish. Some of their books went through several editions and circulated for a long time in Mexico and abroad. It might be expected, therefore, that the indictment would be durable as well. Both elements of the binary formulation elaborated by the nineteenth-century travellers along the Veracruz-Jalapa road, the luxuriant nature and the inadequate inhabitants, might therefore be expected to have filtered through the intervening literature on lowland tropics and influenced the planners of development, the investors, and the in-migrants themselves.

They are, in fact, recognizable in the rhetoric of the frontier spun over Mexico's Gulf lowlands in the 1960s and 1970s. Planners and promoters served up the bounty of permanent streams and rich soils in coloured brochures as fanciful as any nineteenth-century description. Care would have to be taken, of course, and just how much care became evident as time went on, but the immense potential remained intact. The "underdevelopment" of the Mexican tropical lowlands, and similar lowlands elsewhere, was still virtually axiomatic, as was the advisability of applying methods and machines—perhaps with tropical adaptations but often not—from temperate lands already fully modernized. Little attention was given to traditional skills and knowledge, except perhaps as fallback expedients. Most new settlers in the governmentally sponsored projects came from outside of the tropical lowlands and were not familiar with the traditional tropical subsistence strategies. Many had to learn them grudgingly from indigenous or long-resident neighbours when the planned commercial production proved impractical. The objective of agents and clients alike was always something more modern. Even where "development" touched indigenous people, as in the Chontalpa region of Tabasco, there was little reference made by planners to traditional skills and knowledge, and for that matter, little awareness of the actual wants and needs of the clientele. So it was not difficult to see parallels in the judgments of tropical lowlands levelled during the second half of the twentieth century with those expressed in the decades after Mexican independence, the time of Humboldt or, indeed, as far back as the time of Herodotus. The people of the lowland tropics, whether indigenous or in-migrant, could still not be considered equal to the challenge of their environment.

It would be going too far to posit an actual causal relationship between nineteenth-century observations and the decisions made by investors, planners, and settlers in the actual "development" of these lowlands. However, the relationship can certainly be hypothesized. Actually tracking the ideas and their effects might be an absorbing undertaking, which cannot be attempted here.

It is interesting to test for the persistence of the predisposition, as one might use a coring tool to test for the extent of a particular stratum of sediments below the surface of a soil. The geographer Pierre Gourou's book *The Tropical World* comes to mind. It was first published in French in 1946 and translated into English in 1953. It went through three more English editions, the last of which appeared in 1980. It was translated into various other languages and was reprinted at least two dozen times. The authority of the book is repeatedly confirmed in reviews. By 1953 it was already well known, "one of the best on the subject" (Russell 1953:488). In 1967 it was considered, "still *the* book" on the humid tropics (Kirchherr 1967:342). It was called "unique" (PRC 1967:225), and "definitive" (Rinier 1968:291). A reviewer of the English edition of 1980 spoke of Gourou as "the most influential writer to have shaped the Western geographer's understanding of the tropics" (de Souza 1982:57). His book will have been a standard source for anyone in the Western world after mid-century seriously concerned with the development of tropical lands, and it is difficult to imagine that it did not influence planners of tropical development.

Gourou's book treated the tropical world both systematically, that is, with regard to its basic resources and economic activities, and regionally, dealing in turn with its American, African, and Asiatic components. The concluding chapter, "Prospects for the Tropical World," observes regarding Amazonia, but apparently intending a wider application, that "the population is affected by a remarkable incapacity for production" (1966:167). How might this be explained? "[Amazonia's] economy languishes not because the natural conditions enforce this, but by reason of the disastrous historical record" (1966:169). This is an understanding, even sympathetic, view of the incapacity but leaves the indictment intact. Gourou's prescription: "The main prospect for the hot, wet lands of the tropics is the expansion of agriculture over wide areas at present unused" (1966:171). And traditional agriculture is not what was meant: "However poor may be the methods used, commercial cropping is always preferable to mere gathering" (1966:179). Gourou was to change some of his thinking about the tropics in the 1980s, as might be expected. However, those earlier, widely disseminated ideas will have had their effects. In the 1960s and 1970s the meagre use of tropical land was still of concern. Development could not be expected to succeed with traditional methods, especially not shifting cultivation, which was considered extremely wasteful. External agricultural models were still being sought. The people of the tropics were being encouraged to do better with what they had. Were we able to reassemble our nineteenth-century authors and quiet their convivial reminiscence long enough to allow Gourou to lay such thoughts before them, there would be nods of affirmation all around.

The agricultural economist Michael Nelson has analysed a wide range of tropical lowland development projects undertaken according to extraneous development models (1973). His is a fairly hard-headed book, tending to questions of costs, benefits, and policy. Controversy over the timing, methods, and rates of return vis-à-vis investment in long settled neighbouring uplands is noted but not allowed to undermine the basic assumption of the desirability of such development. The Nelson book has been translated into Spanish recently and still seems to have its currency.

Such testing for the old predisposition easily leads to further reflections. A recent summation of lessons learned regarding the colonization of forested tropical lowlands, and there are others, is highly cautionary (Smith 1981). Where traditional means of dealing with the Amazonian environment have been swept aside and extraneous expedients applied in the interests of commercial agricultural production, it has often been with only the sketchiest of information. The results, particularly in governmentally planned projects, have been high costs for the public purse and for the settlers involved, as well as ecological mayhem.

Planned resettlement and regional development projects in the Mexican lowlands present a similarly unprepossessing picture. Enquiries into the views of the settlers in a series of such projects has yielded the dismal realization that planning officials' understanding of tropical resources, of what is feasible and desirable, is often not very much more incisive than the understanding of the foreign travellers passing through the lowlands of Central Veracruz in the early nineteenth century. Moreover, it is also clear that the wants and needs of the people for whom development has been planned have seldom been taken into consideration (Gates and Gates 1976).

A glaring recent example may be found in the huge Uxpanapa resettlement and agricultural development project undertaken in the Isthmus of Tehuantepec (Ewell and Poleman 1980). Again the insights already won regarding the precariousness of a forested tropical lowland environment were largely disregarded. Commercial crop production was introduced precipitously on a vast scale and at a very high cost—with unfortunate results. The remedial application of traditional subsistence strategies, especially shifting cultivation, were disallowed. The settlers' resistance to what they considered counterproductive labour was interpreted as irresponsibility and indeed as laziness; they had become "cumberers of the ground."

It is urgent now to abrogate the historical judgment on the resources and the people of the tropical lowlands. The limitations and vulnerability of the tropical forest ecosystem are well known. It is fortunate that there is also a kind of salvage agro-ethnography combined with experimentation going on in Mexico and elsewhere. Old wisdom regarding the management of tropical resources is being rescued and made functional in the modern context.

Such initiatives involve relearning the exploitation of biological complexity, appreciating old juxtapositions in aid of, say, soil conservation or insect control. The kitchen garden, for instance, has been found to be characteristically quite complex, once mid-latitude, Western ideas on orderliness and rectangularity in gardens are eased a little and the wide range of useful plants that actually grow around most rural tropical houses is recognized. Shifting cultivation has come to be appreciated as more productive, potentially, than it had generally been judged in the past, and more complex. Land subject to seasonal flooding is used throughout the lowlands for dry season pasture or mechanized dry season agriculture. It is known, as was pointed out in earlier chapters, that much of this land was once used for labour-intensive agriculture by means of an ancient system of planting platforms and canals. It may be useful to reactivate this system. At the very least, it would be an instructive outdoor museum piece. The orchestration of these various types of land use and their integration with other subsistence activities, such as fishing, hunting, and gathering, is particularly intriguing.

In actual field investigative practice, shrewdness has probably come to be assumed much more often than laziness. The *campesino*, the man of the land, probably knows very well what he is doing; the imperative is to understand the parameters of his decision-making. This may be quite unwarranted in individual cases, but it has proved a generally productive approach. Sartorius was able to imply a grudging appreciation of the *jarocho's* savoir-faire. It is important to arrive at respect.

References

PRIMARY SOURCES

Altamirano, Ignacio Manuel
   1964 In *Viajes en México: crónicas mexicanas*, edited by Xavier Tavera Alfaro, pp. 295–344. México DF: Secretaría de Obras Públicas
Ampère, Jean-Jacques Antoine
   1964 In *Viajes en México: crónicas extranjeras*, edited by Margo Glantz, pp. 409–29. México DF: Secretaría de Obras Públicas (1856)
Ballentine, George
   1853 *Autobiography of an English soldier in the United States Army*. London: Hurst and Blackett
Becher, Carl Christian
   1834 *Mexico in den ereignissvollen Jahren 1832 und 1833*. Hamburg: Perthe
Beltrami, J.C.
   1964 In *Viajes en México: crónicas exranjeras*, edited by Margo Glantz, 173-240. México DF: Secretaría de Obras Públicas
Biart, Lucien
   1959 *La tierra templada*. México DF: Editorial Jus
   1962 *La tierra caliente: escenas de la vida mexicana, 1849-1862*. México DF: Editorial Jus
Billings, Eliza Allen
   1851 *The Female Volunteer, or The Life and Wonderful Adventures of Miss Eliza Allen, a Young Lady of Eastport, Maine*
Blanchard, Pharamond
   1839 *San Juan de Ulúa*. Paris: Chez Gide
Brackett, Albert Gallatin
   1854 *General Lane's Brigade in Central Mexico*. Cincinatti: H.W. Derby

Bullock, William

1825 *Six Months' Residence and Travels in Mexico; Containing Remarks on the Present State of New Spain, Its Natural Productions, State of Society, Manufacturers, Trade, Agriculture and Antiquities.* London: Murray

1971 *Six Months' Residence and Travels in Mexico; Containing Remarks on the Present State of New Spain, Its Natural Productions, State of Society, Manufactures, Trade, Agriculture and Antiquities.* London: Murray

Bullock, William H. (William Henry Bullock Hall)

1866 *Across Mexico in 1864-5.* London: Macmillan

Burkart, J.

1836 *Aufenthalt und Reisen in Mexico in den Jahren 1825 bis 1834.* Stuttgart: Schweizerbart'sche

Calderón de la Barca, Frances Erskine

1843 *Life in Mexico during a Residence of Two Years in that Country.* London: Chapman and Hall

Coggeshall, George

1858 *Thirty-Six Voyages to Various Parts of the World, Made between the Years 1799 and 1841* (3rd ed.). New York: Putnam

de Fossey, Mathieu

1964 In *Viajes en México: crónicas extranjeras*, edited by Margo Glantz, 241-72. México DF: Secretaría de Obras Públicas (1857)

Domenech, Manuel

1922 *México tal cual es: 1866.* Querétaro: Demetrio Contreras

Elton, J.F.

1867 *With the French in Mexico* . London: Chapman & Hall

Ferry, Gabriel [Louis de Bellemare]

1856 *Vagabond Life in Mexico.* New York: Harper

Fisher, Howard T. and Marion Hall Fisher, editors

1966 *Life in Mexico: The Letters of Fanny Calderón de la Barca.* Garden City, NY: Doubleday & Company

Gardiner, C. Harvey, editor

1959 *Mexico: 1825-1828; the Journal and Correspondence of Edward Thornton Tayloe.* Chapel Hill, NC: University of North Carolina Press

Gilliam, Albert M.

1847 *Travels in Mexico during the Years 1843 and 1844.* Aberdeen: Clark

Grant, U.S.

1883 *Personal Memoirs of U.S. Grant* (2 vols.) New York: Welston & Co.

1952 *Personal Memoirs of U.S. Grant*, edited by E.B. Long. New York: World Publishing

Hall, William Henry Bullock

See William H. Bullock (pseudonym)

Heller, Carl Bartholomaeus

1853 *Reisen in Mexiko in den Jahren 1845-1848.* Leipzig: Wilhelm Engelmann

Henry, Robert Selph
1950 *The Story of the Mexican War*. New York: Bobbs Merrill
Humboldt, Alexander von
1911 *Political Essay on the Kingdom of New Spain* (translated from the
original French by John Black). London: Longman
1809–14 *Versuch, über den Politischen Zustand des Königreichs Neu-Spa-
nien*, Vols. I-V, in 2 vols. (I:1809-V:1814; each paginated separately:
Tübingen: Cotta'sche Buchhandlung
1972 *Political Essay on the Kingdom of New Spain*, edited with an introduction by Mary Maples Dunn (John Black translation [abridged]).
New York: Knopf
1811 *Mexico-Atlas*. Stuttgart: Brockhaus
Jamieson, Milton
1849 *Journal and Notes of a Campaign in Mexico*. Cincinnati: Ben
Franklin Printing House
Kenly, John Reese
1873 *Memoirs of a Maryland Volunteer: War with Mexico in the Years 1846-
7-8*. Philadelphia: Lippincott
Kollonitz, Paula Gräfin
1867 *Eine Reise nach Mexiko im Jahre 1864*. Vienna: Gerold
Koppe, Carl Wilhelm
1835 *Briefe in die Heimat*. Stuttgart: Windenmann
1837 *Mexikanische Zustände aus den Jahren 1830 bis 1832*. 2 vols. Stuttgart and Augsburg: Cotta, Stuttgart and Tübingen
Koppe, Carlos Guillermo
1955 *Cartas a la Patria: Dos Cartas Alemanas sobre El México de 1830*,
trans. and introd. by Juan A. Ortega y Medina. México DF: Imprenta
Universitaria
Latrobe, Charles Joseph
1836 *The Rambler in Mexico: 1834*. London: Seeley and Burnside
McWhiney, Grady and Sue McWhiney
1969 *To Mexico with Taylor and Scott, 1845-1847*. Waltham, MA: Blais-
dell
Maissin, Eugène
[1839] 1961 *The French in Mexico and Texas (1838-1839)*. Paris: Salado,
Texas: Anson Jones Press
Mason, R.H.
1852 *Pictures of Life in Mexico*. London: Smith Elder & Company
Maury, Dabney Herndon
1894 *Recollections of a Virginian in the Mexican, Indian and Civil Wars*.
London: S. Low; New York: Scribner's
Mayer, Brantz
1844 *Mexico as It Was and as It Is*. London: Wiley and Putnam; New
York: Winchester

214   References

Mühlenpfordt, Eduard
   1844 *Versuch einer getreuen Schilderung der Republik Mejico*(2 vols.).
   Hanover: C.F. Kius
   1969 *Versuch einer getreuen Schilderung der Republik Mejico*. Graz:
   Akademische Druck- und Verlagsastalt
von Müller, Baron J.W.
   1864 *Reisen in den Vereinigten Staaten, Canada und Mexico*. Leipzig:
   Brockhaus (1874:35)
Norman, Benjamin Moore
   1845 *Rambles by Land and Water, or Notes of Travel in Cuba and Mexico*.
   New York: Paine and Burgess
Pattie, James Ohio
   1833 *The Personal Narrative of James O. Pattie of Kentucky*, edited by
   Timothy Flint. Cincinnati: E.H. Flint
Payno, Manuel
   1964 In *viajes en México: crónicas mexicanas* edited by Xavier Tavera
   Alfaro, pp. 47-134. México DF: Secretaría de Obras Públicas
Poinsett, Joel Roberts
   1824 *Notes on Mexico Made in the Autumn of 1822*. Philadelphia: Carey
   and Lea
Pourade, Richard F. (ed.)
   1970 *The Sign of the Eagle: A View of Mexico, 1830-1855. Letters of Lt.
   John James Peck, U.S. Soldier in the Mexican War*. San Diego: Union-
   Tribune
Richthofen, Emil Karl Heinrich von
   1854 *Die äusseren und inneren politischen Zustände der Republik Mexico
   seit deren Unabhängigkeit bis auf die neueste Zeit*. Berlin: Hertz
Robertson, William Parish
   1853 *A Visit to Mexico by the West India Islands, Yucatan, and United
   States, with Observations and Adventures on the Way* (2 vols.). London:
   Simpkin, Marshall
Ruxton, George Frederick
   1855 *Adventures in Mexico and the Rocky Mountains*. New York: Harper
Sartorius, Carl
   [1858] 1961 *Mexico about 1850*. Stuttgart: Brockhaus (first published
   in German)
Schiede, C.J.W.
   1828–30 Botanische Berichte aus Mexico. Erster Bericht über die
   Vegetation um Veracruz und über die Reise von dort nach Jalapa.
   Linnaea 4, 205-12
   1828 Zweiter Bericht über die Gegend um Jalapa und Excursion auf
   den Vocan Orizaba. *Linnaea 4*, 212-36
   1829 Dritter Bericht über die Gegenden von Papantla und Misantla

und über die Reise von Jalapa dorthin und zurück, *Linnaea* 4, 554-82
1830 Vierter Bericht. Excursionen in der Gegend von Jalapa und Reise von dort nach Mexico. *Linnaea* 4, 463-77.

Scott, Lt.-General Winfield
1864 *Memoirs of Lt.-General Scott, LL.D., Written by Himself.* New York: Sheldon

Sealsfield, Charles
1974 *Der Virey und die Aristokraten oder Mexico im Jahre 1812: Drei Teile in zwei Bände.* Hildesheim & New York: Olms Verlag

Smith, George Winston, and Charles Judah (eds.)
1968 *Chronicles of the Gringos: The U.S. Army in the Mexican War, 1846-1848: Accounts of Eyewitnesses and Combatants.* Albuquerque: University of New Mexico Press

Taylor, Bayard
1850 *Eldorado, or Adventures in the Path of Empire.* London: R. Bentley; New York: G.P. Putnam

Thompson, Waddy
1846 *Recollections of Mexico.* New York: Wiley and Putnam

Tudor, Henry
1834 *Narrative of a Tour in North America, Comprising Mexico, the Mines of Real del Monte, the United States, and British Colonies, with an Excursion to Cuba, in a Series of Letters Written in the Years 1831-2.* London: J. Duncan

Tylor, Edward B.
1861 *Anahuac.* London: Green, Longman and Roberts

Valois, Alfred D.
1861 *Mexique, Havane et Guatemala: Notes de voyage.* Paris: Dentu

Vigneaux, Ernest de
1964 In *Viajes en México: crónicas extranjeras,* edited by Margo Glantz, pp. 459-98. México DF: Secretaría de Obras Públicas

Wappäus, Johann Eduard
1863 *Geographie und Statistik von Mexiko und Centralamerika.* Leipzig: Hinrichs

Ward, H.G.
1828 *Mexico in 1827.* London: Henry Colburn

Wilcox, Cadmus M.
1892 *History of the Mexican War.* Washington, DC: Church News

Williams, T. Harry
1956 *With Beauregard in Mexico: The Mexican War Reminiscences of P.G.T. Beauregard.* Baton Rouge: Louisiana State University Press

Wilson, Robert Anderson
1855 *Mexico and Its Religion.* New York: Harper
1856 *Mexico: Its Peasants and Its Priests.* New York: Harper

SECONDARY SOURCES

Ackerknecht, Erwin H.
1955 George Foster, Alexander von Humboldt, and ethnology. *Isis* 46:83-95
Adams, Richard E.W.
1977 *Prehistoric Mesoamerica*. Boston and Toronto: Little, Brown
Alatas, Syed Hussein
1977 *The Myth of the Lazy Native*. London: Frank Cass
Allgemeine Deutsche Biographie
1896 Leipzig: Verlag von Duncker & Humblot
Archer, Christon I.
1971 The key to the kingdom: the defense of Veracruz, 1780-1810. *The Americas* XXVII: 4, April:426-49
Armillas, P.
1971 Gardens on swamps. *Science* 174:653-61
Arreola, Daniel D.
1982 Nineteenth-century townscapes of eastern Mexico. *Geographical Review* 72:1-19
Bates, Henry Walter
[1863] 1962 *The Naturalist on the River Amazon*. Berkeley and Los Angeles: University of California Press
Barker, Nancy Nicholas
1979 *French Experience in Mexico, 1821-1861: A History of Constant Misunderstanding* Chapel Hill, NC: University of North Carolina Press
Bermúdez, G. Gilberto
1977 *Jalapa en el siglo XVI*. Tesis maestro en historia, Universidad Veracruzana, Xalapa
Bernal, Ignacio, and Eusebio Davalos Hurtado
1952–53 Huastecos, Totonacos y sus vecinos. *Revista mexicana de estudios antropológicos* XIII
Blok, Anton
1972 The Peasant and the Brigand: social banditry reconsidered. *Comparative Studies in Society and History* 14a;494-505
Bloomfield, Arthur L.
1958 *A Bibliography of Internal Medicine: Communicable Diseases*. Chicago: University of Chicago Press
Boyer, Richard E.
1972 La ciudad Mexicana: Perspectivas de estudio en el siglo XIX. *Historia Mexicana* XXII:2. 142-59
Brennan, Robert D.
1984 Long-distance movement of goods in the Mesoamerican formative and classic. *American Antiquity* 49:1, 27-43

References 217

Bushnell, David, and Neill Macaulay

1988 *The Emergence of Latin America in the Nineteenth Century*. New York: Oxford University Press

Calderón Quijano, J.A.

1953 *Historia de las fortificaciones en Nueva España*. Sevilla: Publicaciones de la Escuela de Estudios Hispano-Americanos de Sevilla (1966)

Calcott, Wilfrid Hardy

1936 *Santa Anna*. Norman: University of Oklahoma Press

Carrera Stampa, Manuel

1953 Las ferias novohispanas. *Historia Mexicana* II (3): 319-42

Carter, Henry R.

1931 *Yellow Fever: An Epidemiological and Historical Study of Its Place of Origin*. Baltimore: Williams and Wilkins

Chappe d'Auteroche, Jean

[1778] 1973 *A Voyage to California to Observe the Transit of Venus*. Surrey, Eng.: Richmond

Cheetham, Sir Nicolas

1970 *A History of Mexico*. London: Rupert Hart-Davis

Clark, Kenneth

1976 *Landscape into Art*. New York: Harper and Row

Compact Edition, *Oxford English Dictionary* 1971

New York: Oxford University Press

Conklin, Harold C.

1968 Ethnography. *International Encyclopedia of the Social Sciences* 5:172-78. New York: Macmillan and Free Press

Dasmann, Raymond F., John P. Milton and Peter H. Freeman

1973 *Ecological Principles for Economic Development*. London: John Wiley & Sons

Deleon, Arnoldo

1983 *They Called Them Greasers: Anglo Attitudes toward Mexicans in Texas, 1821-1900*. Austin: University of Texas

Diaz, Lilia (ed.)

1974 *Versión francesa de México: Informes económicas 1851-1867*. Tlateolco, México DF: Secretaría de Relaciones Exteriores

Diaz del Castillo, Bernal

1939 *Historia Verdadera de la Conquista de la Nueva España* (Vol. 3). México DF: Editorial Pedro Roberto

Donkin, R.A.

1979 *Agricultural Terracing in the Aboriginal New World*. Viking Fund Publication in Anthropology, 56. Tuscon: University of Arizona Press

Duffy, John

1966 *Sword of Pestilence: The New Orleans Yellow Fever Epidemic of 1853*. Baton Rouge: Louisiana State University Press

Durán, Fray Diego
1964 *The Aztecs: The History of the Indies of New Spain*. New York: Orion Press

Eisely, Loren
1961 *Darwin's Century*. Garden City, NY: Doubleday/Anchor

Emerson, Edwin
1912 Mexican bandits at a close view. *The Independent* LXXII (3322), 1 August. New York

Encyclopaedia Britannica
1977 Volume VIII:692 (14th ed.). Chicago: Encyclopaedia Britannica

Estrada y Zenea, I.
1872 *Álbum Veracruzano: Colección de Vistas Fotográficas*. Veracruz: El Progreso

Ewell, Peter T. and Thomas T. Poleman
1980 *Uxpanapa Reacomodo y Desarrollo Agrícola en el Trópico Mexicano*. Xalapa, Veracruz: Instituto Nacional de Investigaciones Sobre Recursos Bióticos

Fleming, Peter
1983 *One's Company: A Journey to China in 1933*. Harmondsworth, Eng.: Penguin

Flexner, James Thomas
1970 *Nineteenth-century American Painting*. New York: Putnam

Florescano, Enrique
1976 Mexico. In *Latin America: A Guide to Economic History 1830-1930*, edited by Roberto Cortes Conde and Stanley J. Stein, pp. 435–543. Los Angeles: University of California Press

Flores Mata, G., J. Jimenez, X. Madrigal, F. Moncayo, and F. Takaki
1971 *Tipos de vegetación de la república mexicana*. México DF np

Flores Salinas, Berta
1964 *México Visto por Algunos de sus Viajeros*. México DF: Ediciones Botas

Florescano, Enrique and Isabel Gil Sánchez (eds.)
1976 *Descripciones Económicas Regionales de Nueva España: Provincias del Centro, Sureste y Sur, 1766-1827*: México DF: Sepinah

Forman, Sylvia Helen
1976 The future value of the 'verticality' concept: implications and possible applications in the Andes, *Actes du XLII Congrès International des Americanistes*, Vol. 4, pp. 233–56. Paris

Fussell, Paul
1980 *Abroad: British Literary Travelling between the Wars*. Oxford: Oxford University Press

García, Enriqueta
1970 Los climas del estado de Veracruz. *Anuario del Instituto Biológico*, UNAM 41, serie botánica (1):3–42

Gates, Marilyn and Gary Gates
1976 Projectismo: the ethics of organized change. *Antipode* 8:3, 72–82
Gibson, Charles
1971 *The Black Legend: Anti-Spanish Attitudes in the Old World and the New.* New York: Knopf
Glacken, Clarence
1967 *Traces on the Rhodian Shore: Nature and Culture in Western Thought from Ancient Times to the End of the Eighteenth Century.* Berkeley and Los Angeles: University of California Press
1970/1973 Man Against Nature: An Outmoded Concept. In *The Environmental Crisis* by Harold W. Helfrich, Jr., pp. 127–42. New Haven and London: Yale University Press
Glantz, Margo (editor)
1964 *Viajes en México: crónicas extranjeras.* México DF: Secretaría de Obras Públicas
Gliessman, S.R., R. Garcia E., and M. Amador A.
1981 The ecological basis for the application of traditional agricultural technology in the management of tropical agroecosystems. *Agro-Ecosystems*, 7 (1981):173–85
Gómez-Pompa, Arturo
1978 An old answer to the future. *Masingira* 5:np. Oxford: Pergamon Press
González de Cossio, Francisco
1958 *Un Plano Desconocido del Pueblo de la Grande Feria de Xalapa, Año de 1776.* Mexico: Los Talleres Gráficos de la Nación
Gourou, Pierre
1966 *The Tropical World: Its Social and Economic Conditions and Its Future Status.* New York: Wiley
Greene, Graham
1950 *Journey without Maps.* London: Heinemann
Grimm, Jacob and Wilhelm Grimm
1942 *Deutsches Wörterbuch.* Leipzig: S. Hirzel Verlag
Gunn, Drewey Wayne
1974 *Mexico in American and British letters.* Metuchen, NJ: Scarecrow Press
Harris, David R.
1972 The origins of agriculture in the tropics. *American Scientist* 60(2):180–93
Hartshorne, Richard
1939 *The Nature of Geography.* Lancaster, PA: Association of American Geographers
Hayward Gallery
1973 *Salvator Rosa.* London: Arts Council

Herrera Canales, Inés
1977 *El Comercio Exterior de México 1821-1875*. México DF: El Colegio de México
Hesse-Wartegg, Ernst von
1890 *Mexico*. Wien and Olmütz: Hölzel
Hobsbawm, E.J.
1969 *Bandits*. London: Weidenfeld and Nicolson
Hofstadter, Richard, William Miller, and Daniel Aaron
1967 *The United States: The History of The Republic*. Englewood Cliffs, NJ: Prentice-Hall
Humboldt, Alexander von
1847 *Kosmos*. Stuttgart: Cotta'sche Verlag
1852 *Personal Narrative of Travels to the Equinoctal Regions of America during the Years 1749-1804*. Vol. 1. London: G. Routledge & Sons
1943 *Südamerikanische Reise*. Berlin: Safari-Verlag
Iglesias, José María
1966 *Acayucán en 1831*. Colección Suma Veracruzana. México, DF: Editorial Citlaltepetl
Immerwahr, Raymond
1972 *Romantisch: Genese und Tradition einer Denkform*. Respublica Literaria 7. Frankfurt: Athenäum
Janzen, Daniel H.
1975 *Ecology of Plants in the Tropics*. London: Edward Arnold
Johnson, William Weber
1966 A lady of manners and mischief. *New York Times Book Review*, 23 October, pp. 36, 38
Jones, W.T.
1961 *The Romantic Syndrome: Toward New Methods in Cultural Anthropology and History of Ideas*. The Hague: Martinus Nijhoff
Kirchherr, Eugene C.
1967 Review of *The Tropical World* (1966) by Pierre Gourou. *The Journal of Geography*, LXVI:342
Krupp, Marcus A. and Milton J. Chatton
1977 *Current Medical Diagnosis & Treatment 1977*. Los Altos: Large Medical Publications
Lameiras, Brigitte B. de
1973 *Indios de México y Viajeros Extranjeros: Siglo XIX*. SepSententas 74, México
Leonard, Irving A.
1972 *Colonial Travellers in Latin America*. New York: Knopf
Liceaga, Eduardo
1910 Annual report on yellow fever in the Mexican Republic, from August 16, 1908, to date. *American Journal of Public Hygiene*, I:63-70

References 221

Lowenthal, David, and Hugh C. Prince
1965 English landscape tastes. *The Geographical Review* IV:186-222
Lowie, Robert H.
1937 *The History of Ethnological Theory*. New York: Rhinehart
Mares, José Fuentes
1964 *Poinsett: Historia de una gran intriga*. México DF: Editorial Jus
Martínez del Río, Pablo
1952-53 Elogio de Jalapa. In *Huastecos, Totonacos y sus Vecinos* edited by Ignacio Bernal and Eusibio Davalos Hurtado, pp. 19-25. Mexico: Sociedad Mexicana de Antropología
Mayer, William
1961 *Early Travellers in Mexico, 1543-1816*. México DF: Published by author
Medellín Zeníl, Alfonso
1979 Clásico tardío en el centro de Veracruz. *Cuadernos Antropológicos* 2:205-13
Meir, N.C.
1963 Review of Jones' *Romantic Syndrome*. In *Contemporary Psychology* April:160
Melgarejo Vivanco, José Luis
1960 *Breve historia de Veracruz*. Xalapa: Universidad Veracruzana
1979 *Los Jarochos*. Xalapa: gobierno del estado de Veracruz
1979 Personal communications
Mentz de Boege, Brígida Margarita von
1980 *México en el siglo XIX visto par los alemanes*. México DF: Universidad Nacional Autónoma de México
Mentz de Boege, Brígida, von and Verena Radkau, Beatriz Scharrer, and Guillermo Turner
1982 *Los Pioneros del Imperialismo Alemán en México*. México DF: Ediciones de la Casa Chata
Miranda, F., and A.J. Sharp
1950 Characteristics of vegetation in certain temperate regions of eastern Mexico. *Ecology* 31:313-33
Miranda, José
1962 *Humboldt y México*. Mexico: UNAM
Murra, John V.
1956 *The Economic Organization of the Inca State*. PH.D. thesis in anthropology, University of Chicago
1978 *La Organización Económica del Estado Inca*, México DF: Siglo Veintiuno
Nebel, Carl
1839 *Viaje pintoresco y arqueológico sobre la parte más interesante de la Republica Mejicana, en los años transcurridos desde 1829 hasta 1834*. Paris and Mexico: Renouard

Nelson, Michael
   1973 *The Development of Tropical Lands: Policy Issues in Latin America.*
   Baltimore and London: Johns Hopkins University Press
Newton, Norman
   1969 *Thomas Gage in Spanish America.* London: Faber & Faber
Ninth Pacific Science Congress
   1958 Climate, vegetation, and rational land utilization in the humid
   tropics. In *Proceedings of the Ninth Pacific Science Congress of the Pacific
   Science Association,* Vol. 20. Bangkok: Secretariat
Ochoa Villagómez, Ignacio
   1885 *Vegetación espontánea y repoblación de los medanos de la zona litoral
   de Veracruz*
Oeste de Bopp, Marrianne
   1979 Die Deutschen in Mexico. *Die Deutschen in Lateinamerika: Schick-
   sal und Leistung,* edited by Hartmut Fröschle, pp. 475-564. Tübingen
   and Basel: Horst Erdmann Verlag
O'Gorman, Edmundo
   1961 *The Invention of America.* Bloomington: Indiana University Press
Ortega y Medina, Juan Antonio
   1955 *México en la Conciencia Anglosajona.* México DF: Gráfica Pan-
   americana
   1960 *Humboldt desde México.* México DF: UNAM
P.R.C.
   1967 Review of *The Tropical World* (1966) by Pierre Gourou. *Geography*
   52:225
Palerm, Ángel
   1953 Etnografía antigua totonaca en el oriente de Mexico. *Revista
   Mexicana de Estudios Antropológicos* XIII:163-73
Parsons, James J.
   1972 Spread of African pasture grasses to the American tropics. *Journal of Range Management* XXV:12-17
Pasquel, Leonardo
   1969 *Biografía Integral de la Ciudad de Veracruz: 1519-1969.* Colección
   Suma Veracruzana. México DF: Editorial Citlaltepetl
   1970 *Aspectos de la navegación mexicana.* México DF: Editorial Citlaltepetl
Pasquel, Leonardo (editor)
   1979 *Viajeros en el estado de Veracruz.* México DF: Editorial Citlaltepetl
Paz, Octavio
   1961 *The Labyrinth of Solitude.* New York: Grove Press
Peckham, Morse
   1970 *The Triumph of Romanticism.* Columbia, SC: University of South
   Carolina Press

Pelzer, Karl J.
1958 Land utilization in the humid tropics: agriculture. In *Proceedings of the Ninth Pacific Science Congress of the Pacific Science Association 1957*, vol. 20, 124–43. Bangkok: Secretariat
Pferdekamp, Wilhelm
1958 *Auf Humboldts Spuren*. Munich: Max Hueber Verlag
Praz, Mario
1962 Review of Jones' *Romantic Syndrome*. *Modern Language Review* October:585
Prescott, William H.
1843 *History of the Conquest of Mexico*. London: Routledge and Sons
Prévost, M., and Roman d'Amat
1951 *Dictionnaire de la biographie française* (5th vol.). Paris: Librairie Letouzey et Ané Sàrl
Prieto, Guillermo
1968 *Una Excursión a Jalapa en 1875*. México DF: Editorial Citlaltepetl
Puleston, Dennis E.
1977 The art and archaeology of hydraulic agriculture in the Maya lowlands. In *Social Process in Maya Prehistory: Studies in Memory of Sir Eric Thompson*, edited by N. Hammond, 449–64. London: Academic Press
Ratzel, Friedrich
[1878] 1969 *Aus Mexico: Reiseskizzen aus den Jahren 1874 und 1875*. Stuttgart: F.A. Brockhaus
Real Academia Española
1956 *Diccionario de la lengua española*. Madrid
Rees, Peter
1976 *Transportes y comercio entre México y Veracruz, 1519-1910*. México DF: SepSetentas
Revista Jarocha
1968 Número dedicado a María Enriqueta Camarillo de Pereyra. March (53). México DF: Editorial Citlaltepetl
Rich, Daniel Cotton
1946 *Henri Rousseau*. New York: Museum of Modern Art
Richards, P.W.
1952 *The Tropical Rain Forest: An Ecological Study*. Cambridge, Eng.: Cambridge University Press
Rinier, James A.
1968 Review of *The Tropical World* (1966) by Pierre Gourou. *The Professional Geographer* 20:291
Rivera Cambas, Manuel.
1959–60 *Historia Antigua y Moderna de Xalapa y de Las Revoluciones del Estado de Veracruz* Vols. I-XVII (1839–43). México DF: Editorial Citlaltepetl

Robinson, Cecil
1977 *Mexico and the Hispanic Southwest in American Literature.* Tucson:
University of Arizona Press
Rodríguez, Joaquin María
1971 *Apuntes sobre el canton de Xalapa.* México DF: Editorial Citlaltepetl
Russell, E. John
1953 Review of *The Tropical World* (1953) by Pierre Gourou. *International Affairs* 29:488
Rzedowski, Jerzy
1978 *Vegetación de México.* México DF: Editorial Limusa
Sancholuz, Luis Alberto
1984 *Land Degradation in Mexican Maize Fields.* Interdisciplinary PH.D.
thesis, University of British Columbia
Sanders, William T.
1952-53 The anthropogeography of Central Veracruz. *Revista Mexicana de Estudios Antropológicos* XIII:27-78
Sanders, William T. and Barbara J. Price
1968 *Mesoamerica: The Evolution of a Civilization.* New York: Random
House
Santamaría, Francisco J.
1959 *Diccionario de Mejicanismos.* México DF: Editorial Porrúa
Sauer, Carl
1971 The formative years of Ratzel in the United States. *The Annals of
the Association of American Geographers* 61(2): 245-54
Secretaría de Recursos Hidráulicos
1971 *Boletín Hidrológico* (43). Región hidrológica no. 28 Parcial, ríos
Actopan, La Antigua y Jamapa
Siemens, Alfred H. and Lutz Brinckmann
1976 El sur de Veracruz a finales del siglo XVII: un análisis de la
"relación" de corral. *Historia Mexicana,* XXVI:2, 263-324
Siemens, Alfred H.
1980 Indicios de aprovechamiento agrícola prehispanico de tierras
Inundables en el Centro de Veracruz. *Biótica* 5:3, 83-92
1983 Wetland agriculture in pre-hispanic Mesoamerica, *Geographical
Review* 73:2, 166-81
Silva Herzog, Jesús
1944 *Relaciones estadísticas de Nueva España de principios del siglo XIX.*
México: Secretaría de Hacienda y Crédito Público
Singer, Charles, and E. Ashworth Underwood
1962 *A Short History of Medicine.* Oxford: Clarendon Press
Smith, Ashbel
1951 *Yellow Fever in Galveston, Republic of Texas, 1839.* Austin: Univer-
sity of Texas Press (1951 edition includes a biographical sketch by

Chauncey D. Leake)
Smith, George
1917 *The Dictionary of National Biography*. London: Oxford University Press
Smith, Harold F.
1968 *American Travellers Abroad: A Bibliography of Accounts Published Before 1900*. Carbondale: Southern Illinois University
Smith, Nigel J.H.
1981 Colonization lessons from a tropical forest. *Science* 214:755–65
de Souza, Anthony R.
1982 Review of *The Tropical World* (1980) by Pierre Gourou. *Tijdschrift voor Economie en Soc. Geografie* 73:57
Stafford, Barbara Maria
1984 *Voyage into Substance*, Cambridge, MA: MIT Press
Stamp, L. Dudley
1958 An approach to rational land utilization (summary and conclusions). In *Proceedings of the Ninth Pacific Science Congress of the Pacific Science Association 1957*, Vol. 20, 161–69. Bangkok: Secretariat
Stein, Stanley J. and Barbara H. Stein
1970 *The Colonial Heritage of Latin America: Essays on Economic Dependence in Perspective*. New York: Oxford University Press
Swift, Jonathan
1963–65 *The Correspondence of Jonathan Swift*, edited by Harold Herbert Williams (5 vols.). Oxford: Clarendon Press
Tavera Alfaro, Xavier, ed.
1964 *Viajes en México: Crónicas mexicanas*. México DF: Secretaría de Obras Públicas
Thümmel, A.R.
1848 *Mexiko und die Mexikaner, in Physischer, Socialer und Politischer Beziehung*. Erlangen: Palmische Verlagsbuchhandlung
Time-Life
1974 *The Expressmen*. Old West Series. New York
Tomikichiro, Tokurichi
1963 *Tokaido*. Osaka: Hokusha Publishing
Tomlinson, H.M.
1912 *The Sea and the Jungle*. New York: Random House
Trens, Manuel B.
1955 *Historia de la H. Ciudad de Veracruz y de su Ayuntamiento*. México DF: Talleres Gráficos de la Nación
Troll, Carl
1959 *Die Tropischen Gebirge*. Bonn: Dümmlers Verlag
Vallier, Dora
1964 *Henri Rousseau*. New York: Abrams

Vásquez Yañes, Carlos
1971 La vegetación de la Laguna de Mandinga, Veracruz. *Anales del Instituto de Biología*, Serie Botánica 42:49–94
Vigneaux, Ernest de
1979 In *Viajeros en el Estado de Veracruz*, edited by Leonardo Pasquel, pp. 245–59. Mexico: Editorial Citlaltepetl
Villaseñor y Sánchez and D. Joseph Antonio
1746 *Teatro americano*
Vivó Escoto, Jorge A.
1964 Weather and climate of Mexico and Central America. In *Handbook of Middle American Indians*, Vol. 1, edited by Robert C. West, pp. 187–215. Austin: University of Texas Press
Wagner, Philip L.
1964 Natural vegetation of Middle America. In *Handbook of Middle American Indians* 1:216–64
Wallace, Alfred Russel
1889 *A Narrative of Travels on the Amazon and Rio Negro*. London: Ward, Lock and Co.
Wanklyn, Harriet
1961 *Friedrich Ratzel: a biographical memoir and bibliography*. Cambridge, Eng.: Cambridge University Press
Ward, James S.
1972 *Yellow Fever in Latin America: A Geographical Study*. University of Liverpool: Centre for Latin-American Studies
Warren, Andrew J.
1951 Land-marks in the conquest of yellow fever. In *Yellow Fever*, edited by George K. Strode, pp. 1–37. New York: McGraw-Hill
Wehr, Hans
1971 *A Dictionary of Modern Written Arabic*. Wiesbaden: Otto Harrassowitz
West, Robert C., and John P. Augelli
1966 *Middle America: Its Lands and Peoples*. Englewood Cliffs, NJ: Prentice-Hall
Wilhelmy, H.
1966 Tropische Transhumance: Beobachtungen auf einer Amazonasreise mit G. Pfeifer und H. Lehmann. In *Geographische Arbeiten 15*, 198–207. Wiesbaden: Heidelberger
Wolcott, Roger (ed.)
1925 *The correspondence of William Hickling Prescott, 1833–1847*. Boston: Houghton Mifflin
Zolá Báez, Manuel
1980 *Estudio de la vegetación de los alrededores de Xalapa, Veracruz*. BA thesis in biology. Xalapa: Universidad Veracruzana

Index

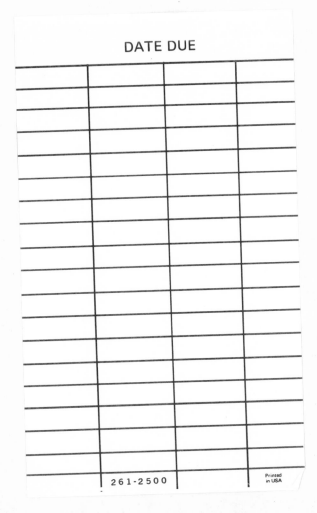

DATE DUE

261-2500

Printed in USA